What, Then, Is Time?

Also by Eva Brann

The World of the Imagination: Sum and Substance

What, Then, Is Time?

Eva Brann

ROWMAN & LITTLEFIELD PUBLISHERS, INC.
Lanham • Boulder • New York • Oxford

ROWMAN & LITTLEFIELD PUBLISHERS, INC.

Published in the United States of America
by Rowman & Littlefield Publishers, Inc.
4720 Boston Way, Lanham, Maryland 20706
www.rowmanlittlefield.com

12 Hid's Copse Road
Cumnor Hill, Oxford OX2 9JJ, England

British Cataloging in Publication Information Available

Library of Congress Cataloging-in-Publication Data

Brann, Eva T. H.
What, then, is time? / Eva Brann.
p. cm.
Includes bibliographical references and index
ISBN 0-8476-9292-2 (cloth : alk. paper)
1. Time. I. Title.
BD638.B73 1999
115—dc21 99-19014
CIP

Printed in the United States of America

∞ ™ The paper used in this publication meets the minimum requirements of American National Standard for Information Sciences—Permanence of Paper for Printed Library Materials, ANSI/NISO Z.39.48–1992.

To Christopher and Adrian Sangiamo,
born December 1995 and June 1997

Contents

Preface

What, Then, Is Time?—the title of this book and its guiding question—is also part of a quotation: "What, then, is time? If nobody asks me, I know; if I want to explain it to him who asks, I don't know." It comes from Augustine's *Confessions* (XI 14), and makes its obligatory appearance in most books on time. To judge from my own case, the reason is that it rings the bell of a common experience.

INCITEMENTS

Augustine's observation was not only a first motive for this inquiry, but a first frame for it. Why don't I know what that is which I tell, save, spend, mark, waste, and even kill every day of my life with perfect aplomb? It must be that time is peculiarly elusive, something always there that nonetheless evades our grasp. But "something" is wrong; time can't, for all its power over our lives, be an entity. In fact, my first title for this book was "The Potent Nonentity," until my friends Gretchen and Barry Mazur helped me find one that was both simpler and more telling. And though I never lost the sense that I was looking for something at once pervasive and insubstantial, I saw that Augustine's very way of bringing on the question was itself an incitement to reflection. Was part of the perplexity caused by the questioner's abruptness? Is the nature of time perhaps the sort of subject that calls for us to insinuate and weasel ourselves into its depths along numerous paths? This book makes its approach to the question in just that way.

There was, however, also a circumstantial motive for taking up the question of time. My previous preoccupation had been with the imagination (*The World of the Imagination: Sum and Substance*, 1991). It turned out that to think about the imagination entailed thinking about its treasure-house and workshop, that is, memory. But memory is largely (though not altogether) of the past, and the past is a phase of time. It was a confirmatory pleasure to realize that Augustine himself had composed the book on memory (X) as a prelude to the book on time (XI).

There were yet more incitements to thinking about time after studying the imagination. The images of the imagination and the structures of time have curious logical, or better ontological, similarities. An image, as a likeness, is composed of Nonbeing and Being *at once*, meaning that it *is not* the original, which in a way it also *is*; an image is the presence of an absence. In time, as the pure structure of Becoming, that "at once" comes apart as absence turns into presence and presence into absence, as the future that *is not* yet ceaselessly propels the present that *is* now into a past that *is not* anymore; time is thus a present winged by two absences. I will try to make sense of these formal affinities in Part Two.

There too I will talk about certain time-pathologies, particularly the nightmarish sense of being powerless before an oncoming future. And again I had found that the imagination is prone to similar capitulations to the ghost of futures to come. In fact, these first incited me to wonder whether or not time was a power. And finally it seemed to me, the more I thought about it, that such reflection had moral import, that coming to terms with time might, even must, influence the conduct of life—more than some specifically ethical disquisitions I have read.

INTENTIONS

Even the modes of thinking about time and imagination seem to me to be similarly located in the topography of consciousness: There is the flash-by of decision making exercised in the external world, which is—as I know from seven years' officiating—the sleepwalking of the intellect. At the other end of the cognitive spectrum, there is the pure absorption in beings of thought and their relations, for which the intellect is fully awake but, at least now and then, out of the world and beyond even our personal interior. And then there is the middle ground, our sensibility, where is to be found both the space of the imagination and the sense of time. This is the place of our intimate, personal interiority, the center of what used to be called the soul. It is to be visited in the mode of introspective reflection conjoined both with experience and with thought. In this book neither the multifarious social and literary theories of time, nor the huge and useful fields of time-measurement and time-management, nor the ingenious constructions of formal time-logics will be taken up. My main intention is to study and reflect on the inner experience of time and to look for its roots, its ground. Though the language of that search cannot help but be that of the inquiring intellect, that is to say, of philosophy, its object, which appears to be neither a thing nor an abstraction, is as centrally human as anything can be.

Finally, in one, perhaps only apparently incidental, aspect time and imagination are practically inverses of each other—in the quantity of philosophical literature they have induced. The imagination is rich in features, and the prime reason is no doubt that it takes place in that quasi field called inner space, so that from the purest geometric structures to the most sense-informed image formation it is full of articulable determinations. Time, on the contrary, is thin and property-poor and rich only in perplexities. Hence a number of philosophers have observed that there can be no proper science of a nonentity so attenuated in nature that it must borrow its representation from one of the dimensions of a quasi entity like space.

As it happens, the number of extended philosophical treatments of both these enigmas, time and imagination, are each in inverse proportion to their subject's endowments: huge for time, minuscule for the imagination—so small that in the writing of *The World of Imagination* I came to think of the imagination as the "missing mystery" of philosophy and therefore made the treatment very long. By similar reasoning its complement—*What, Then, Is Time?*—is intentionally much shorter.

READERS

I thought the following kinds of readers might be interested in this book:

- People who want to learn about time by finding their way into fundamental theories that are perennial, diverse, deep, and originative
- Aficionados of this inexhaustible subject who acquire, as I have done for a decade, every book that shows promise of taking up their perplexities
- Students who, by choice or imposition, need to come to grips with the primary works on time
- Teachers who might wish to base a course on the choices and selections I have made (usually small parts of works) and to take advantage of the interpretative expositions here given for exercising their students' critical powers.

I can even imagine someone taking this book to the beach to have at hand while meditating among the sands on the mysteries of time.

ORGANIZATION

This book is conceived in two unequal parts. The longer first part, called "Presentations," contains studies of selected texts by those authors who seem to me to have written most deeply and most engagingly about time. Since study is long and invention short the second part, called "Reflections," which contains my own thoughts on the question that is the title of this book, is much briefer.

Of Part One, only the first chapter presents a thesis: that external or natural time is (almost) always a name for something discernibly different, namely change. Implied in this critique of natural "time" (the scare-quotes are from Einstein) is a first division of the field of time as a subject into *outside* and *inside* time. Making this division and mounting this critique has the convenience that external time, which is taxonomically one-half and in content far more than half of the enormous subject-territory of time, is thereby sliced away. The division itself has the warrant of general acceptance. Almost everyone agrees that only more-than-merely conscious beings, that is to say, self-conscious beings that have a self-acknowledged interiority, in short *human* beings, *have* past and future. Divinity is beyond time, while stones and stars, dust mites and even dolphins in their lovable congeniality are only *in* time, in *our* time: They may *be* clocks but they *have* none. The assumption here, to be supported in Part Two, is that the so-called "phases," past, present, and future, are deeper features of time than is mere continual or continuous succession.

In the rest of the chapters, on internal or psychic time, I have let the texts carry me whither they would, while I present explications of immediate meaning and wider significance without more than this or that marginal note of doubt or admiration. The chief philosophical texts on time vary from hard to hellishly hard, so my main effort was to clarify, without forgetting that clarifications are always interpretations. I have tried, keeping the possible impossibility of the undertaking well before me, to figure out what each text, taken as the expression of an authorial intellect belonging to the same genus as my own, meant to convey. For it seemed to me that the point of study is to get the good of another's thought, and that, wise hermeneutic inhibitions notwithstanding, it must be possible for me to grasp what was written for me to read.

I also avoided quarreling with alternative interpretations, because that undertaking requires a double space, one to present the other view fairly and another to argue with it in the nuanced way it deserves; so I refrained, both because I was under size restrictions and because it was less my object to enforce my particular opinion than to help to keep the question alive.

Early on I saw what seemed to me a remarkable fact, and it became the organizing principle of Part One. Pairs of authors far apart in time and even further apart in the philosophical framework within which they presented their understanding of time seemed to have theories whose nuclear thought was similar to the point of identity. This observation, if corroborated, had two effects I welcomed. The incidental one was to reinforce my suspicion that ways of thinking about time are not time-bound. The main effect was to give shape to four ways of understanding time, four crux-theories, so to speak, that anyone who wants to draw nearer to the truth of this matter has to grapple with. Time is

- A clock or what it shows
- A counted pulsing of the soul
- A self-elongation of the soul
- A stretching of the mind.

The first four chapters are, then, devoted to elucidating these provisionally mystifying characterizations after extracting them carefully from their very disparate philosophical frameworks. I have done my best to give an abbreviated but unskewed rendition of these environments—just enough to site each theory of time and to make its terms intelligible.

Below is a table of the pairs of authors and the textual selections on which Part One is focused. The last two pairs are of authors who are antithetical rather than similar in their theories, but who proved to be in some way pivotal to the presentation.

Chapter One

Plato, *Timaeus* 37–38 (before 347 B.C.)	Einstein, "On the Electrodynamics of Moving Bodies," para. 1 (1905)

Chapter Two

Aristotle, *Physics* IV 217 ff. (after 335 B.C.)	Kant, *Critique of Pure Reason,* Transcendental Aesthetic and Logic (1781, 1786)

Chapter Three

Plotinus, *Ennead* III 7 (c. A.D. 270)	Heidegger, *Being and Time,* para. 46 ff. (1927)

Chapter Four

Augustine, *Confessions* XI (A.D. 401)	Husserl, *Phenomenology of Internal Time-Consciousness* (1905)

∞ ∞ ∞

Chapters Four and One

Newton, *Principia*, Scholium to Definitions (1686)	Leibniz, *Metaphysical Foundations of Mathematics* (after 1714)

Chapter One

Hegel, *Philosophy of Nature* 257–61 (1817)	Bergson, *Time and Freewill,* chapter II (1884)

REFERENCES

The reader will notice, whether with delight or dismay, that this book has neither footnotes nor long bibliographies.

I have absorbed the references to primary works right into the text. Citations of secondary works are acknowledgments of particular help actually incorporated into my text. Works so used are referenced in one amalgamated bibliography. I must, however, confess that after years of reading on this huge subject, I could not always tell which of the ideas and facts in the stockpot of my memory had come to me by study and which by independent discovery.

The primary texts, which are the main objects of study for Part One, are always cited by the paragraph or page numbers that are standard over all editions and translations. To send readers back to the text and to help them test my presentations with ease, textual references are copious.

Concerning general bibliographies, it is an open secret that these descend from book to book with individual variations, so they are here omitted as redundant. Almost every work of scholarship referenced contains good specific lists. Samuel Marcey's *Time: A Bibliographic Guide* of 1991 gives a—daunting—overview of the time-related publications, estimated at 180,000, that have appeared since 1900. At that rate, there must by now be over 20,000 more, including the effort before you: Time is an inexhaustible incitement to inquiry.

Finally, I take pleasure in thanking my colleague Peter Kalkavage, who read the last draft with that discerning care both for the subject itself and for my thought about it that helped me to say what I meant to say and to see what I might have said, Ann Martin, who prepared the manuscript with keen-eyed competence, and Timothy Martin, who did the diagrams.

Eva Brann
St. John's College
Annapolis, 1998

Part One

Presentations

Chapter One

External Time: Time as Motion

This first chapter is by way of a clearing of the decks. Some things about time are not, upon reflection, believable, no matter how often they are said and how obvious they begin to seem. The questions concerning time are, as I will have ample occasion to show, elusive to formulate, the answers gather perplexity as they gain in profundity, and the number of treatments becomes almost as confounding as the subject itself. To discern through this coruscating haze even one clearing of certainty, even if it be merely negative, is a comfort and not a bad starting point.

Here is what seems clear to me: When time is spoken of as being in the external world, the world of nature primarily, but also of society and history, it is usually a word for something else—for motions of various kinds and for their measurements. To put the point plainly now, before it gets very complicated: When time is named in natural science, particularly physics, what is meant is a standard motion or a probabilistic tendency, and most physicists have the grace to eschew definitions of time. Again, when it is spoken of in anthropology, for example, when the Mayan "concept of time" is presented, the object of study is an intricate and significance-bearing calendrical system, and I have not come across the claim that the Mayans even had a separate word for the notion of time (León-Portilla, 17); they seem to have used the word for "sun." When time is casually spoken of with respect to history, as it is so often that it does not bear telling, anything at all might be meant, from a dry date to a significant epoch, from mere passages to driving forces, and nowhere is the use of the word more protean. Time in history is for that reason beyond my inquiry here.

I shall try to say under the two headings that comprise this chapter exactly what it might mean to say that time, when used in respect to the external world, means something else, for this claim loses its point in the absence of some precision. It should be said right now, however, that there *are* exceptions, though only two that I know of, one theological and the other philosophical. The very strongest of the theological arguments for real external time is made by Isaac Newton, who posits an absolute, equable flux that is truly antecedent to, and unmeasurable by, any motion and that has a very strong theological implication: God is in one aspect temporal (p. 153). I will argue later that Augustine's internal time also requires external Creation-time on which it, so to speak, rides, but his doctrine is not as singular or forceful as Newton's (p. 118).

The other exception is the time developed out of space by G. W. F. Hegel in his *Philosophy of Nature*, which will be set out in section B of this chapter. This time too is antecedent to matter and motion, and is therefore real, pure, external time. In general, nothing throws more light on what writers mean by time than how they relate it to space. That is why section B is devoted to (1) a particularly deep consideration of this question, that by Hegel, which, as I say, develops time from space and turns out to be indispensable for isolating the notion of pure external time, that is, of time that is not surreptitiously motion, and (2) an unusually passionate defense of time as being a deeper and humanly more serious notion than space, that by Henri Bergson, which set the tone for many subsequent treatments of time both as an external process and as an interior human dimension.

But first, in section A, I present two authors whose treatment of time amounts to a philosophically candid and practically useful acknowledgment of its nonbeing in the cosmos or the universe.

A. PLATO AND EINSTEIN: TIME AS A CLOCK

In the rest of this book I will be thinking about passages, pages, chapters, and whole books. But the first thesis about time will be presentable in a mere sentence or two, taken from authors just a millennium and a half apart.

I. Plato: The Cosmic Clock

> He thought of making a certain movable image of eternity, and, at once with ordering heaven, he made an eternal image going according to number, that which we have named Time. (37 d 6 ff.)

This quotation comes from the dialogue *Timaeus* and is, I believe, the second most often quoted sentence on the nature of time in the literature—just second to Augustine's sentence, which will appear so often in this book, on the difficulty of speaking about time. The Timaean passage is not only the precipitating definition for Plotinus's theory of time (p. 96) and for all his Neoplatonic successors (Sambursky and Pines, 9 ff.). It is also a favorite quotation for time-romantics and physicists alike, for the former because of its cosmic grandeur and for the latter because it says what they think: that the cosmological frame of the world is its fundamental clock.

Before analyzing its sense, it must be said that neither Plato nor Socrates is speaking this sentence, but Timaeus, an astronomer (27 a 3) and a person of uncertain existence, whom Plato has made to share some opinions of the Pythagoreans, among which was the notion that time is the revolution of the embracing sphere or even of the whole heaven itself. It hardly matters, however, whose views are behind those of Timaeus, as long as they are not taken to represent Plato's theory of time. The dialogues never make time thematic (Brann 1994, 466 f.). There are, to be sure, early versions of still-debated time-puzzles concerning the "phases" of time, namely "is," "was," and "will be," and the moment of temporal transition, the "sudden instant" (*Parmenides* 141, 151 e ff.; p. 177). There is also the myth of reversed time, the Age of Cronos, the father of Zeus (whose name is often identified with *Chronos*, Time), an age in which human beings are earth-born in the prime of life, nurtured by the earth and ruled by the god; there follows the age of unwinding when humans are human-born and toil and govern themselves while the epoch declines into chaos—until the god's time takes hold again (*Statesman* 268 d ff.). But there is, as I said, no extended treatment of time in the Platonic dialogues, and though the Timaean myth cannot be regarded as a *logos*, a rational account of time, it may supply the reason why Plato has offered none.

For Timaeus's "likely story" or, more literally, "like myth" (*eikos mythos*) is presented as a feast to Socrates in return for the account of the best city and its citizens that he had given the day before. Timaeus is going to provide a cosmic container for this polity. And just as Socrates had in the conversation of the previous day, a conversation that is either identical with or close to the one recounted in the *Republic*, founded political discourse, so Timaeus will provide the mode of discourse proper to cosmology and mathematical physics in general, the myth or story that is at once likely *and* like.—It is, in short, what we call a model. Timaeus is not only presenting a model of the cosmos but also supplying the reason why such a model can plausibly be a likeness of the real heavens: because the heavens are themselves an image modeled on an aboriginal paradigm, rationally and beautifully designed by an artificer-god, who has himself made the great heavenly image. Here is the first account of model making in physics, showing the two defining features that this activity still has today: It frankly produces nothing more (or less) than plausibly rational artifacts, and they have the character of mathematical images.

And that is why the story of the origin and nature of time is not the Socratic, and therefore not a central Platonic, theory. Socrates, and Plato behind him, account for becoming in its instability and for the appearances in their variety by the "hypotheses" of the forms, which make stabilizing and unifying language possible and are the ground of Being and the cause to things. The chief Platonic problem is just how the world of appearances participates in the world of forms. The Timaean "cosmos," on the other hand, is, as the Greek word signifies, already a place of beautiful order, as good as possible, a place of signifying shapes and numbers, and it originates by the work of said divine artificer, the demiurge, who "orders heaven" (*diakosmon*), making a "movable image" (*kineton eikon*) by imitating within the stuff of the world of becoming a model (*paradeigma*) of being. If one is bent on speaking of a Timaean "theory" of time, then one should say that what is there presented is Plato's theory of astronomical-physical time. It will be part of the thesis of this chapter that world-time and soul-time, external and internal time, are usually given quite disparate origins in our tradition, a tradition that begins exactly here.

Yet that claim leaves one marvelous circumstance out of account. If time is the numerable motion of the heavens, it might indeed be thought to be the most exterior thing imaginable. But the outermost embracing frame of Timaeus's cosmos is made of soul-material. It constitutes what we call the celestial equator and the ecliptic circle that is oblique to it. In this encompassing psychic frame, Being or the Same, and its Other or Nonbeing, are blended in numerical ratios and then shaped into the circles that contain Becoming. The demiurge makes these cosmic arrangements so that the soul may not be junior to the body but may be its ruler (34 c ff.). That is to say, Plato acknowledges in his cosmic story that temporality outside of a numerally disposed soul is practically unimaginable. It will be shown over and over to be in effect unthinkable.—Time needs to be imagined and thought as psychically contained. Of course, it is from this mythic vision of intrapsychic world-time that Plotinus's ontological derivation of time will take off.

Now to a closer look at the quotation. Time is an *image* of eternity. That means it has the dual image-features of being at once *like* and also *not* the original. The difference-feature is expressed in a pun: The word for eternity, *aion*, sounds like the privative of *ion*, the word for "going." Eternity is the stable "ungoing"; its movable image goes, and it goes "according to number." That is as much as to say that the heavens are a cosmic clock. Timaeus says that the god designed the sun, the moon, and the other five stars, together called the "planets" (literally, the "wanderers"), and placed them in their circular orbits so that time might have its generation (*genesis*), for from them comes the definition and guardianship of the numbers of time; thus time is distinguished into its parts, the days and nights, months and years—the units of time. In another place Timaeus speaks of the stars as the "instruments of times" (*organa chronon*, 41 e 5, similarly of the earth and moon, 42 d 4). The heavenly bodies are the indicators or, one might say, the moving dial, of the clock. (Although the Greeks had no circular analog-dial clocks, they were well acquainted with circular measurement of time through the chief diurnal clock they did possess, the sundial, whose upright shadow-throwing indicator was, significantly, called the *gnomon*, the "knower.")

The similarity of the heavenly clock to its eternal paradigm is, of course, the steadiness of its circular celestial motion, a motion of which Aristotle will show in the *Physics* that it is the most stable, continuous, and regular motion there is (264 b 9 ff.). The celestial motion is *standard* motion.

As time has parts marked by the orbital motions of the instruments of time, so time generates two forms (*eide*): "the was" and "the will be." These apply properly to the two "motions" of time and the becoming that "goes on" within time (38 a 2). But we, Timaeus says, use speech wrongly when we apply these tenses of the verb "to be" to Eternal Being; they belong only to Becoming and the things that are borne about in the world of sense. He says that his discourse is not the opportune moment (*kairos*) for "precise speaking" about time, and that is, I imagine, why he omits the grave consequences of the converse of his prohibition: If the tensed forms of "to be" are not to be used of Being, the tenseless form "is" is not to be used of Becoming. Herein lies one of the most serious, if not the most serious perplexities of time: the impossibility of speaking rightly of the human present, which is enmeshed in Becoming. In any case, Timaeus implies that the "phases" of time—past, present, future—arise with human speech about time, and they will indeed henceforth be human-centered.

Here is a crucial point: Time is *not* natural passage (Taylor 1928, 680). The heavenly bodies are mobiles and move, and their going gives rise to the epiphenomenon of time and its numerable parts. It is *we* who name this moving image Time. What can be seen here at its origin is that baptism of motion as time that befits physics and bedevils philosophy. It is Timaeus who performs it.

II. Einstein: The Local Clock

> It might appear possible to overcome all the difficulties attending the definition of "time" by substituting "the position of the small hand of my watch" for "time." (I, 1)

What is said in the intimation-fraught originating myth of Timaeus is repeated a millennium and a half later with stripped-down candor by Albert Einstein, only that the Timaean clock is cosmic, or better, that the cosmos *is* the clock, while the Einsteinian clock needs to be local and is an artificial device. The quotation is from Einstein's first presentation of the Special Theory of Relativity in his 1905 paper "On the Electrodynamics of Moving Bodies." Together with his little popular book *Relativity* (1916), this paper is by far the most thoughtful and accessible presentation of the theory to be found. One sign of Einstein's care is the fact that he regularly puts quotation marks around the word "time." We are in the hands of a physicist who knows exactly what he is about; in all the textbooks of physics that I know, the meaning of time is set aside with much more nonchalance (p. 13).

It is in fact possible to substitute the clock's reading for time only if one can provide for the synchronicity of clocks very far apart in space. In the Timaean cosmos, and in the later physicists' universe, time was, one way or another, understood as having the feature of worldwide simultaneity. That is to say, people considered it intelligible and necessary that it should be "the same time" in the most distant parts of place or space. Since Einstein is not preoccupied with the nature of time, but rather with the time of nature, or to put it precisely, since he is not, as physicist, interested in the Being of time but in its measurement, simultaneity becomes problematic for him. For there is "no mystic communion of clocks" (Diane Ackerman in Sobel, 34). It is often said that the Special Theory of Relativity is a theory of measurement. Certainly it begins very simply with a reflection on the measurement of time—though it ends in mind-boggling new notions of the nature of time. But careful philosophers of science never lose sight of the fact that these temporal consequences are based on a certain theory of time-measurement, that is to say of observers and their time-instruments.

Einstein begins with the clock that is free of the problem of simultaneity, the clock that measures the time *here*, right by the observer. His understanding of clock-time or watch-time is operational, that is to say, the quantity measured is time if it shows up on identically constructed, similarly operating instruments, which instruments are *assumed* to go "*at the same rate*" (*Relativity*, ch. VIII). It is apparently of no interest in this context that there is something circular in the notion of instruments going "at the same rate," for a rate of motion, or a velocity, is itself a ratio of distance to time. The reason the circle is of no importance is that faith is (implicitly) placed in a standard motion, an equable passage that underlies time, just as was done in the *Timaeus*. Once again it should be said that this faith is perfectly sufficient for physics, though rousingly problematic for philosophy. It is this difficulty Einstein seems to be mindful of in his scare-quotes about time.

In converting universal simultaneity into localized synchronicity, Einstein begins by filling space with a grid of clocks as defined above: One universal time is replaced by innumerable clocks and their observers. Some imaginative physics texts actually depict parts of this grid (Taylor and Wheeler, 18, fig. 9). He then *defines* synchronicity in the most commonsensical way available without making unwarranted assumptions. Since in a theory of time-measurement in which readings are to be taken at great distances, the speed-characteristic of the signal is all-important, and since light has the greatest, and as it turns out, the most constant speed, Einstein uses light signals and *defines* the time taken by a signal to travel from observer A and his clock to distant observer B and his clock to be equal to the time for its reflection from B back to A. In other words, the "time" required by light to travel from A to B is to *mean* the same quantity as the "time" from B to A. Only thus, Einstein says, does the definition of a "common time" become possible, that is to say, only thus can meaning be given to clocks running synchronously. We usually think of clocks that are not very far apart as synchronized when they show a common time for the same event, such as the arrival of a light signal. Why having a "common time" should be defined as light taking equal time coming and going from clock to clock is, however, although soon to become a crucial test, quite unperspicuous until a further assumption is added to the definition.

But before going on, Einstein says that synchronization is symmetrical and transitive: If a clock at A is synchronous with a clock at B, the converse also holds, and if two clocks are each synchronous with a third, then they synchronize with each other. In other words, synchronous clocks can be stationed all over. He also says at this point that "time" is now defined: It means *the* time of an event, which is given by the stationary clock located at the place of the event and synchronized with a stationary reference clock. The starting wish to say that time is what the clock shows has been justified for all clocks stationary relative to each other.

The assumption that Einstein next introduces "in agreement with experience" is that the mean velocity of light for *any* round-trip is a constant, which is equivalent to saying that the velocity of light is a universal constant. It is worth noting that the experimental terrestrial measurement of the speed of light, depending as it does on the reflection of light, uses Einstein's "equal times" definition as an assumption; it is the mean velocity that is calculated. This assumption makes the definition real, that is, applicable to nature. For observers will now know how to use it in practice to synchronize clocks: Supposing A sets his clock at 0 when he sends the signal, B needs only to calculate how much time light, traveling at its known velocity, would need to cover the (known) distance from A to B and set his clock to exactly that time as soon as A's signal arrives. When the signal is reflected back to A, both observers can rest assured that the definition of synchronicity has been realized. A's clock will show a time equal to that needed for light at its known speed to cover twice the distance AB, for light in fact takes the same time to travel from A to B and from B to A. B's clock will show exactly the same, though the directions are reversed: Both clocks will show $t = 2AB/c$ (where c is the speed of light) at the return of the signal to A. And that means they are synchronized in the ordinary sense.

So far everything is as it always was except for a grand shift to theoretical candor: Einstein replaces a tacitly assumed simultaneity of all times everywhere in space by an explicitly defined synchronicity of clocks at indefinitely many locations. The notion that time is really a standard motion and that times are given by its natural or artificial units is what physicists from Timaeus on rarely acknowledge. The new elements in the first paragraph of Einstein's 1905 paper are (1) the circumstantial recognition of the willful fact and the careful consideration of what its logic requires, and (2) the inclusion within theory of the fact that, clocks being instruments of measurement, the observers' relation to them is of the essence. But, of course, the Special Theory of Relativity itself, in which physical time will undergo such remarkable conceptual changes, has not yet come on the scene.

An obvious—and our usual—way to synchronize clocks is to bring them together, to set them to read the same time, and then to move one to the intended distant location. This method is not available to Einstein, because it assumes that moving clocks are not affected in their timekeeping. This assumption turns out to be wrong, and that is why Einstein has to define synchronicity for clocks that are stationary relative to each other and already at their intended location. It turns out, however, that clocks that seemed synchronous to observers in the stationary system are no longer so to observers in systems in uniform motion. The criterion of "equal times" will be violated, roughly speaking, because the (uniform) velocity of the observer's system will enter differently into the equation for the times coming and going. Thus, the notion of a common time is not absolute, but is rather relative to the stationariness or motion of the observer.

At this moment, just when the theory takes off, I shall leave it, since my point was to show how conceptually complex the apparently reductionist notion that time is what the clock tells really is in its full-blown form. Or at least, I will give only the sketchiest rendition of the fate of time in the Special Theory.

For it is well known that the fate of time is to lose its independence. In classical physics time is usually what is called the "independent" variable. That denomination has various senses. It signifies, of course and first of all, that a large number of physical formulas are stated as functions of time, meaning that the time-value is freely chosen, and other physical quantities are then determined by means of a function. It may also serve as a reminder of the reason why time is the freely chosen variable: For any given time, particle positions are uniquely fixed, but the converse does not hold; a position may recur at indefinitely many times. Or, above all, the independence of time may allude to its separateness from and, moreover, its priority to, space, features that ultimately derive from the sense that time progresses, whereas space only sprawls. (For the sense in which time is reversible in classical physics, see p. 14.)

What is meant by saying that time, and so of course space as well, loses its independence? Well, for one thing, time will turn out to have more to lose than space. To give, as I said, the most broad-brushed description of what happens: Einstein lays down two major principles that must be preserved in the theory to follow: (1) The Principle of Relativity requires that the laws of nature appear in exactly the same way in *all* coordinate systems that are in uniform translatory motion relative to each other, or: there is no absolute preferred system. (2) The speed of light is not only uniform in all directions in the stationary system, but remains the same whether emitted by a stationary or a moving body, or: the ratio of light path to time interval equals a constant velocity c under all circumstances.

Taking these principles into account, Einstein works out the transformations that relate observers in stationary and in moving systems, which turn out to be the set of formulas known as the Lorentz transformations. They differ from the so-called Galilean transformations of classical kinematics in this: In the classical transformations the space-coordinate alone was affected in the shift to a moving system, while the time-coordinate remained the same. In the new transformation not only the spatial variable but *also* the time variable is affected. Time, which stood independently outside the older transformation, has become implicated in the spatial translation. The effect is that *each* of two observers in two systems moving relative to each other finds that the space-measuring rod carried by the other is shorter than his own and seems to be subject to a contraction in the direction of motion. And as a complement, *each* finds that the other's clock is running slower than his own, so that each thinks the other's time is stretched in the direction of motion or subject to a dilation in comparison with his own. The amount of space contraction and time dilation depends on the relative velocities of the systems, and is, in fact, apparent only for very large velocities, while for small ones the Galilean and Lorentz transformations are nearly the same. The equations are such that the transformed time-coordinate is gotten not only through the time-coordinate but *also* through the space-coordinate of the original system. Since these are kinematic equations, meaning that no mass is involved, the contractions and dilations are not intrinsic to the bodies involved, the measuring rods and clocks, but are perspectival. They depend on each observer's motions and are reciprocal or relative to each other, in short, *unreal* (Whitrow 1961, 211).

So we end where we began: It is not time we are talking about but the consequences of a theory for the measurement of time together with certain physical assumptions. Time belongs to metaphysics, physics works with "time."

A huge and unending controversy, sometimes referred to as the "Battle of the Gods and the Giants," has arisen concerning the following question: Since Einstein had banished simultaneity from space, and simultaneity is an ingrained feature of time as it relates to space, should time be banished from physics and should the physical world be "spatialized" or "geometricized"? Time in such a model would be entirely privatized, confined to the observer's "proper" time at a certain location. The world would become absolute, "pre-written," and the observer would waft through it from perceptual moment to moment, experiencing as Becoming what was really his particular intersection with Being. This view was encouraged by Hermann Minkowski's geometrization of the Lorentz transformations as rotations of a four-dimensional Cartesian coordinate system that leave certain quadratic expressions invariant. Since in Minkowski's representation one of the axes is a time-axis, which enters into the determination of the space-axes, the space of the diagram is referred to as space-time, though the time coordinate is usually expressed as a spatial dimension and called "time-like." This coordinate system stands for a changeless "absolute" world, through which the observer travels along a path called a "world-line." To make it imaginable in our three-dimensional space, at least one of the purely spatial dimensions is omitted. In the physical interpretation of this diagram, the space breaks up into four regions around the origin, which is the location of any observer who is here and now in this present location. A cone opens above the origin containing all space-time points toward which the observer can send signals and thus represents the future. A cone hangs down from the origin from all points within which light signals may reach the observer, and thus represents the past. The remaining regions, which grow very small as the ratio of the velocity that appears in the transformation equation becomes very small compared to *c* and makes the cones open out, represent a space that has no determinate signaling or causal relation to the cones, and in which notions of past and future are scrambled and time loses its principal feature of succession.

The great debate, in which Einstein himself was an ambiguous bystander, concerns this question: Is the formal geometric possibility of a temporally indeterminate world-region a fact fatal to the mathematical view of time as a linear succession and the physical view of time as being or evidencing or determining a causal order? Does the diagram put an end to the phasal understanding of time, in which the world consists of unchangeably determined past events and at least partially open future events, still to be determined by chance and choice?

The defenders of Becoming (the Giants) against the proponents of Being (the Gods) of a completely "written" world, where all that changes is the observers' perceptual perspective, argue that the "world" of the Minkowski diagram is an extreme abstraction from the real world and is subject to misreading: The non-temporal regions are devoid of physical meaning in just the way that Einstein made simultaneity at a distance meaningless by showing, in the words of A. A. Robb, that "an instant cannot be in two places at once." Just as no absolute metaphysical ground of simultaneity lurks behind the relativization of simultaneity, no absolute geometric world lurks behind the relativization of time and space of uniformly moving systems. Moreover, the observer's travel history, the world-line, never intersects itself or does anything but advance in the old-fashioned mode of temporal succession, so that within the geometric region of physical possibility, time marches on as always.

There will be set out below other arguments for preserving the reality of time in physics (section III). But, of course, the time that is being saved is really "time," that is, motion or a measurement of motion, and the human time that is to be preserved is really observer's "time," the noting of clocks and signals. Hence the debate is at bottom Timaean: how to use the imagination in interpreting mathematical models that seem to be true to that in the universe which presents itself to rationalizing observation. In one element, though, the Giants are very suggestively on the mark—in speaking of time as "Becoming." The relation of Time to Becoming has to be clarified if the problem of external time is to be made clear and distinct (section B). (For the debate see Whitrow 1961, ch. I, 1 and passim; de Beauregard 1981; Čapek 1981; Griffin, passim; Harris 1988.)

The embroilment of time with space in the Special Theory of Relativity arises by the combination of a new view of time-measurement that includes the observer with the empirically supported principles that there is no absolute, privileged coordinate system in physics and that the speed of light is a universal constant. It is therefore a result or consequence of physical-mathematical reasoning. In section B more naive and therefore more fundamental reflections on the implication of time with space will be presented: reflections on a primal externalization of time, on its genesis out of space, and on its rescue from space. But first the very notion of exterior time has to be more closely considered.

III. Time among the Physicists

One object of this chapter is to find a way to frame the question whether, if time is what the clock tells, time is or is not something. Does such a "nothing but" definition, a reductionist understanding of time as what the clock tells, do away with time or does it simply postpone the inquiry into the essence of time while making it operationally available to science? If a carefully specified timing-device is given, be it a clock that incorporates an adjustable gnomon, a pendulum, a coiled spring and escapement, a quartz crystal, or be it a chemical reaction, an atomic oscillation, or whatever yields a regular periodic motion, if a display is provided, be it a dial (analog) or number window (digital) or even the sun and the stars themselves, if units of measurement are established, whether by subdivision of the sidereal, true solar or mean solar day, into hours, minutes, seconds, down to nanoseconds, so that one may read off *the* time or calculate the *span* of time between two readings—if all these conditions are present, has a question been settled or has it been set aside?

Certainly the intention in respectable standard physics textbooks is to take good timing-devices for granted and get on with it, but few writers on physics are so radically careful as Einstein.

> The physicist regards time as that familiar though inscrutable one-dimensional something which separates changes in bodies and individual sensations. . . . (Barton, *Analytic Mechanics*, 1911, quoted in Feather, 38)

The implication is not only that the physicist's time is a one-dimensional, space-like continuum (an assumption that is questioned by physicists who hypothesize time-quanta), but that a mysterious *something*, a something that happens between the noted states of change, is being symbolized. Again, when a more recent textbook says that the measurement of time must be able, as above, to answer both the question "What time is it?" and "How long does it last?" (Resnick and Halliday, 7), the dual property of time as noted Now and measured span is acknowledged, even as the thing itself is bypassed.

More philosophically disposed physicists speculate; Heinrich Herz, for example, in his *Principles of Mechanics* of 1894, intending to establish mechanics on a priori principles, begins in a Kantian vein with time as an internal intuition, and draws from that its independence as a variable in physics (Herz, 45). Richard Feynman, giving his exuberantly opinionated introduction of time in the *Lectures on Physics* (sections 5.2–7), is the fairest example of throwing up one's hands without letting go the thing. He despairs at the circularity of the dictionary definition (a time is a period and a period is a time), nonchalantly throws in a very deep suggestion (time is what happens when nothing else happens, when we are just waiting, p. 206), and passes on to say that what really matters is how we measure time, and that the definition of measurable time is finally in terms of natural periodicity; he leaves the question of the continuity of physical time versus time-quanta, that is, the possibility that there might be a time span (called a "chronon") in principle below measurement, to future decision by physics. I should note here, lest a reader imagine that I care about, or that others who reflect on time care about, nailing down its definition, that that is hardly ever the case. A definition is at best a mnemonic of completed thought and has no interest except as a summary of past philosophic activity or as the starting point of a formal system. The issue is *speculation* about time, of which I am giving examples among physicists.

Ernst Mach, in his *Science of Mechanics* of 1883, presents a forcefully radical critique of Newton's dual time, the ordinary relative time that is the physicist's measured time and beyond it an absolute time, a flux independent of and prior to change, which, although compatible with measured time, is not required by the latter and is not measurable "by comparison with any motion. . . . It is an idle metaphysical speculation" (273). Newton is, in fact, the only physicist who posits an external time genuinely distinguishable from motion (for his own words and the strange theological consequences, that God becomes temporal, see p. 153), and is thus rightly distinguished by Mach.

Mach says that we arrive at the idea of time by connecting what is "contained in the province of memory" with what is in sense-perception. When we speak of time as having a definite direction we mean that physical and therefore *physiological* events generally occur in that manner; all this expresses a "profound connection of things" (275). Thus human time has been naturalized.

Therefore:

> It is utterly beyond our power to *measure* the changes of things by time. Quite the contrary, time is an abstraction, at which we arrive by means of the changes of things. (273)

In other words, time is an epiphenomenon of natural processes. Mach, incidentally, outdoes Aristotle in this naturalization of time, for Aristotle will insist that a counting soul is necessary for motion to be numbered (p. 45). It was this book of Mach that greatly influenced Einstein by "its incorruptible scepticism" (Schilpp, 21). Einstein himself, however, thought that *within* Newton's physics the independent existence of time and space was required to give acceleration meaning (*Relativity*, appendix V).

These candid positivists of time, as opposed to the circumventors, induce a kind of semantic collapse. They pull the supporting base from under common words so that we don't quite know what, if anything, we are talking about when we use them in accustomed ways. But if it is the business of vigorously advancing physics to deny the existence of unmeasurable and therefore useless entities, the business of gropingly reflective philosophy is to pursue whatever beings seem to reveal themselves in human thought and speech and to find the greatest interest in what is apparently most useless, though possibly most significant. And nothing could be a more salutary background to that inquiry than the "incorruptible scepticism" of the more thoughtful physicists regarding the external reality of time.

The greatest impetus to talking of time in nature comes from the discoveries of the various "arrows of time." The background to these discoveries is "reversibility," the fact that in classical physics the laws of nature apply indifferently whether time is run backward or forward (Watanabi, 543–47). Some physicists think of this property of physical laws as making physics timeless in the sense that its laws are indifferent to the direction of time, but this atemporal view is distinct from the critique of the concept of time itself. For "reversing time" means, of course, reversing the path of a moving object. Moreover, it is occasionally noted by physicists that celestial orbits are not *in fact* ever reversed, because that would require a sudden deceleration, that is, a catastrophic application of force. Furthermore, it is also noted that to register reversibility or irreversibility a linearly progressive time, the time of the remembering observer, is needed. So on several counts, reversibility is a problematic notion from this point of view of determining the directionlessness of time.

Be that as it may, in 1927 Sir Arthur Eddington introduced the phrase "the arrow of time," which is now used in at least six contexts.

(1) In thermodynamics the directionality of time is expressed in a number of converging ways, as the dispersion of heat, the running-down of isolated systems, the increase of entropy or disorder, the loss of information.

(2) In cosmology it is seen in the expansion of the universe, that is, in the dispersal of matter; Stephen W. Hawking, in *A Brief History of Time* (ch. IX), relates the cosmological to the thermodynamic arrow and that in turn to the human brain and thus to the directionality of human memory, as Mach had already intimated.

(3) In the behavior of a subatomic particle called the neutral K meson, the arrow appears as the irreversible decay it undergoes.

(4) The advance of electromagnetic waves, such as light, which always travel into the future, evinces the arrow.

(5) In psychology a forward flow of time, an ability to judge its rate (subject to distortion by different physiological and psychological conditions) is observed that is analogous to the arrows of physics (Morris, ch. VIII).

(6) In geological and evolutionary science the notion of arrow-time and cycle-time are found useful; the arrow symbolizes the uniqueness, the cycle the repetitions of evolutionary development (Gould, 184 ff.).

In each arrow, what has direction is really a change: a motion, a distribution, a decay, a process, a development belonging to a body or a system of bodies, a particle or a physical structure, a biological organism or a species. But it is not time.

IV. Fallacies

Neither the inadvertent realism nor the agnostic nonchalance among the natural scientists concerning time seems to me satisfactory. The first is refuted by thoughtful physicists themselves, as we have seen. Moreover, though I might be brought to believe that beings of thought lie behind the appearances, beings not seen except with the eyes of thought, it is plain incredible that there might be an entity *within* nature that never in any way shows its proper self to any of the senses—to the external senses, that is, because there is evidently an internal sense of time. To be sure, this particular incredulity of reason does not extend to those notions of external time that have a theological base, such as Newtonian flux or Augustinian Creation-time—those might be met with a different sort of unbelief.

External physical time therefore seems to me like one of those ubiquitous ghosts that haunt different disciplines as pseudo-entities, omnipresent in word but impotent in deed: the brain in cognitive studies, evolution in ethical theory, society in historical accounts. These are examples of terms that loom large in the beginning of expositions and often turn out to bear no specific explanatory responsibility for the particulars of the discipline. It seems, incidentally, to be the case that whereas in physics time is thought of as present in name but absent in effect, in history, it is often thought of as present both in name and as a force (pp. 183, 192).

It is never easy to make clear to oneself just what is faulty in the handling of such terms, but it is even harder to explain why they enjoy such currency. In the case of time, which is neither a material object nor a collection of individuals, nor even a law-governed process (except by just that strange reverse metonymy I have been decrying, which gives to the concrete event the name of its abstracted property, that is, calls change by the name of time), it is yet harder to explain why all of us feel so familiarly attached to a word whose meaning poses such formidable problems and to the existence of an entity we never experience. The nonchalant agnostics will claim that it does not matter at all what word one uses, as long as the clocks accurately tell—whatever, and as long as people give you the time if you ask for it. They are surely right to decline being distracted from getting on with science; they are probably right in believing that linguistic indifferentism is usually—not always—harmless, but they are just wrong in suppressing their natural wonder at what all of us, in clueless familiarity and baseless trust, continually say.

I shall name two root-fallacies that might explain common opinion, and then give some reasons why an ardent interest in these perplexities rather than preoccupied indifference seems to me indicated.

1. *The Fallacy of Misplaced Concreteness*

The name of this fallacy is borrowed from Whitehead's *Science and the Modern World* (1925, ch. III). It is "the error of mistaking the abstract for the concrete." (It is only fair to mention that he uses the fallacy to argue that the abstractions of time in physics cannot prove its unreality in the world.) It seems to me that there is a natural if unsafe urge in us to suppose that when a word and its symbol keep turning up in respectable contexts they always mean something concrete, full featured, and real—as indeed they usually do. Reference to this fallacy does not, however, explain much unless one can also say how time got into physics to begin with.

Galileo, when he introduced the new science of kinematics (that is, those laws of motion that are formulated without respect to mass) in his *Two New Sciences*, began his treatise with an old definition and a set of axioms so designed as to make it demonstrable that in uniform motion, velocities have the physical dimension of space over time, *s/t* ("Third Day," 148). In his terms, he proves that the velocities of mobiles carried in equable motions have to each other a ratio compounded of the spaces run through and the inverse ratio of the times; in modern terms this means that velocity varies directly as the spaces and inversely as the times. The "old definition" is taken from Aristotle's *Physics* (237 b 27–30); Aristotle says that if we take parts of a motion commensurable with the whole (i.e., divisible by an integral unit common with the whole) the times for the motion can be similarly subdivided into integral units. Galileo generalizes it to hold even when the parts are incommensurable with the whole. Now for Aristotle time is not a physical dimension at all, but a discrete magnitude, a counting-number (p. 42), and that is what Galileo inherits and outstrips; he represents time spatially as a line, a continuous magnitude that may have a ratio to any other line, commensurable or incommensurable. In other words, he makes measurable time homogeneous with space so that he may represent it by a spatial element in his diagrams. Thus at the very beginning of modern science, time enters as a spatialized yet independent physical dimension (though for a brief preceding moment distance had been given that role, Kepler, 53). Therefore I speculate that it was this inscription of time into its more sensually apprehensible counterpart, space, that made it more real to the imagination than it had a right to be; Bergson says something of this sort (p. 31).

What, one might wonder, would thinking about the basic physical dimensions of physics be like if, at the very outset, Galileo had ruthlessly eliminated time and conceived velocity instead as the ratio of the distance covered by any mobile to the distance accomplished by a standard mobile with an equable periodic motion, in other words, by a clock? Kinematics would have been spatialized early, and would, moreover, have been saved from a certain ambiguity that always hangs over its dimensions: On the one hand, as qualitatively different entities they were prohibited by the Euclidean theory of proportions that Galileo inherited from occurring in the same ratio (Euclid, *Elements* V, Definition 3) for the good reason that relations of quantity between qualitatively different things are unintelligible. And even though Galileo is always careful to use only ratios of spaces to spaces and times to times in his proportions, since by Proposition 16 of the Euclidean theory magnitudes in a proportion can be alternated, that is switched, in such a way as to set distance against time (if S:s::T:t, then S:T::s:t), there is a temptation to perform the switch and treat the ratio as a mere number-quotient; that is, in fact, exactly what his velocity proposition eventually comes to mean. On the other hand, once that radical step of making all dimensions homogeneous by abstracting completely from their quality is taken, the science loses much of its intuitive relation to reality (p. 190).

Although I am arguing that time really enters science as a symbolic abstraction from physical processes that is then fallaciously concretized or naturalized, there is another, conspiring reason why people see time in nature.

2. *The Pathetic Fallacy*

This fallacy was delineated by John Ruskin in *Modern Painters* (1856, vol. III, iv, 4 ff.) as a critique of an aesthetic attitude to nature, that of projecting onto natural objects our own passions and investing them with a borrowed pathos, willfully by minor poets and morbidly by people of feeling. I want to appropriate this fallacy for the projection of time into nature that no one escapes. For there is surely an inner sense of time (whose origins and nature remain to be worked out), a sense that has three aspects: an inexorable temporal passage, an accumulated duration, and a continual breaking of time into past and future about a present. It is natural to think that nature has the same experience, is borne on the same flux, knows the same permanences that seem to underlie our consciousness, and that natural objects have at each moment pasts and futures as we do. To be sure, if we try to shift from sympathy, that is, fellow-feeling, to empathy, that is, identification, we are immediately rebuffed and sobered. Although poets and even philosophers have managed it, I cannot impute to a stone the capacity of just waiting for something (which was, recall, Feynman's best guess at the meaning of time). Still, in our daily mode, without much imputing, we take the world to be temporal as we are. It is surely a healthy thing, since it makes us feel familiar and at one with that inside-invested outside we call our world.

Then why make a problem of the matter? For one thing, reflective scientists themselves are not indifferent to the question what in ordered change, or process, is to be called time and whether the strand extracted by thought might survive as having independent reality. To put the question another way: Is time an epiphenomenon, just a surface property of process, or is it rather its deep structure and motor, or even, as "concrete time" in Whitehead's phrase, process itself? Some scientists, like Ilya Prigogine, hope that the reintroduction of process-time into an atemporal and spatialized physics in which time is only a symbol will rehumanize nature by bringing it closer to temporal humanity (Griffin, 16). The issue for them is not only that in process-time they find a new meaning of time and that the rapprochement of human and natural time seems salutary, but that thought about external time may actually influence science itself. But for the nonscientist the attraction of the question lies, I think, in the fact that the resolution shapes our view of ourselves: Are we continuous with nature and natural, or are we, in some aspect, at least, utterly discrete and a-natural—in our temporality? If there is external process-time, we belong to nature; if not, we are, to some extent, outside it.

There is yet another respect in which the real existence of external time should concern us. In the accounts given of that part of the external world that is formed by human activity, particularly in history, the theoretical assumption of an extrapersonal time has huge consequences, effects not only rhetorical but practical—mostly bad, as I hope to show in the second part of this book (p. 184). But there is no coming to grips with that external human time without having come to some conclusion about external natural time. For who can imagine a suprahuman historical time that does not participate in the time of nature?

V. Division of the Field

In focusing on one circumscribed area, external time or time in space, a division of the field has tacitly been made: time in the world, time in the soul. Such a division is indispensable in dealing with a subject whose existence lacks evidence and whose essence is elusive. But is *this* way of cutting into the problem the best one? There are so many possible approaches. One might begin by distinguishing succession-time from phase-time, the time that flows continuously on from the time that is segmented around the present into past and future. Or one might start with the differentiation of time from space and investigate why time can be represented in space and whether some qualitative difference is lost in doing so. Or one might first inquire whether time is an originary being or force or a derivative relation or epiphenomenon. Or one might ask whether time is at bottom permanence and duration or rather transition and flux. Or one might treat it as a logical structure abstracted from its practical effects. All these pairs of alternatives, and quite a few more, are, of course, implicated with each other, but how the first division is made colors all the outcomes. This lesson is to be learned from Plato's *Sophist*, in which the object of investigation is the Sophist, whose nature is as slippery within the field of human expertise in general as time is within the field of human experience as a whole. And just as Socrates divides and redivides the field until the Sophist is finally caught in his most proper subdivision, so too, anyone who wants to get a grasp on time through the study of theories about it eventually has to make an educated guess which theoretical distinctions so divide the field as to pen in what is most nearly Time itself.

The prime division actually employed in this book, that into external and internal time, recommended itself to me for the following reasons. (1) I thought I could exclude one side of the division and so narrow the search by showing that external time is a fallacious notion. (2) The distinction is in any case made by nearly all theorists of time, and they themselves more often than not exclude one or the other branching, either for being, albeit conceivable, irrelevant to their enterprise, or as being falsely conceived to begin with. (3) Human temporality itself may turn out to be the reason why we can think of ourselves in the first instance as interior beings confronted by an exterior world.

1. *Leibniz: Relation of Succession*

Before I present my outline of the divisions in the field of time, one author pivotal in the scheme must be briefly brought on the scene. Gottfried Leibniz delineates an external time, as do the preceding physicists with whom he therefore belongs. Yet this physical time is almost completely denatured, and merely relative in the sense that it has no independent influence on nature. Leibniz is usually coupled with his antithetical twin Newton, who postulates the most absolute—though equally effectless—external time, an underlying equable flux; it turns out to be formally so oddly similar to Husserl's idea of an absolute internal flow that it is treated there (p. 153).

The terse doctrines of Leibniz's *Metaphysical Foundations of Mathematics* (after 1714) are more helpful here than the theological controversies of his famous correspondence with Samuel Clark, in which, it is commonly represented, Leibniz demolishes Newton's notion of absolute time (1715–16).

Leibniz carefully prepares the ground for time as a relation by attributing "prior" and "posterior" to the states of things themselves: "If one of two states which are not simultaneous involves a reason for the other, the former is held to be *prior*, the latter *posterior*." If two states are not in opposition to each other, and neither involves a reason for the other's existence, they coexist, and the order of their coexistence is space. Then

> *Time is the order of existence of those things which are not simultaneous.* Thus time is the universal order of changes when we do not take into consideration the particular kinds of change. (Leibniz 1956, 1083)

Here Leibniz is quite close to Aristotle, who also posits "before and after" as outside of and prior to time, and finds the reason for motion within the mobile itself (p. 41). Time, then, is for Leibniz simply a relation whose magnitude is called duration, a relation that abstracts from the nature of change. The bare formulation that "time is the order of succession" is simply mystifying without the understanding that this succession is already causally ordered, and that time is a mere quantitative abstraction from the successive states. The causes are in the states of things, not in time. Leibniz comes as close as any writer I know to reducing external time to featurelessness, and his time ranks at the bottom of the enumeration by potency of conceivable kinds of external times.

2. *Table of Theories of Time*

Here, on the brink of time becoming internal, a retrospective and prospective table of the types of time that the internal-external classification yields seems in order. The technical terms that appear in this table will be explained under the authors that introduced them.

- I. External Time (in descending order of reality)
  - A. Real and pure time: the absolute flux (Newton).
  - B. Time as a symbol for motion: the independent coordinate (classical physics).
  - C. Time immanent in physical process as directionality: the "arrow of time" (e.g., thermodynamics, cosmology, evolution).
  - D. Time immanent in process as cause: the "passage of the cause into the effect is the cumulative character of time" (Whitehead).
  - E. Operational time: what the clock tells
    1. the cosmos: the diurnal clock (Timaeus).
    2. local clocks: the observer's reading (Einstein).
    3. organismic timing: the biological clock (e.g., circadian rhythms of plants and animals).

F. Calendrical systems: preoccupation with event-cycles and chronology (interpreted as interest in time, e.g., the Mayan calendars).
G. Epochal time: human events driven by superpersonal historical forces (historicism).
H. The abstracted order of succession: relative time (Leibniz).

II. External and Internal Time Coordinated
A. Clock-readings and observers' time-consciousness: the dual aspect of time-measurement (most physicists).
B. Time as the count and the counting of motion: objective and subjective number (Aristotle).
C. Time as pure Becoming arising through the self-differentiation of space: time as the incipient life of space (Hegel).
D. Pure duration as the internal interpenetration of heterogeneous moments and its homogenizing symbolization in space: the intensivity and extensivity of time (Bergson).
E. World-time and soul-time: succession and memory (Augustine).
F. Time as a dual series: an external series of earlier and later event-positions depending for their temporality on a second series of past-present-future event-positions, the latter requiring memory, perception, and inference (McTaggart).

III. Internal Time (as presented by the writers discussed in chapters two through four of Part One)
A. The counting soul: psychic motion (Aristotle).
B. The "train of ideas": mental succession and duration (Locke).
C. The counting soul: the temporal "form of intuition" (Kant).
D. The grounding of time: the "diastatic" temporalization of the soul (Plotinus).
E. The grounding of time: the "ecstatic" temporalization of human Being (Heidegger).
F. The "stretching" of the mind: the intersection of a memory-and-expectation axis with world-time at the point of perception (Augustine).
G. The "stretching" of consciousness: the intersection of a retention-and-protention axis with the internal time-flux at the point of primary impression (Husserl).

B. HEGEL AND BERGSON: TIME OUT OF SPACE

The two writers presented under this heading wrote within the same century and so are chronologically much closer than the other pairs appearing in this book. Yet what they have in common is more on the surface than what divides them: They share a specific interest in the relation of time to space and in explaining the urge, almost universal in the Western tradition, to represent time analogously to space, as a rectilinear dimension. It is this common interest in the involvement, for better or worse, of time with space that places them in this chapter on external time. Neither of them, however—and that makes this section something of an appendix—has the faith in physics as a way into the nature of things that is shared by the preceding thinkers from Plato's Timaeus to Leibniz.

Hegel, however, engages in a dialectical analysis of the genesis of time out of space, to my mind the deepest reflection on external time available. What he succeeds in doing is working out a notion of pure Becoming such as is needed to make intelligible a time that is in nature without being simply identical with motion or its measurable parts.

Bergson does not engage in dialectic concerning time and space but offers a kind of philosophical psychology. Hence, his intention is not to derive one from the other but, on the contrary, to separate space and time from the mistaken entanglement that the human propensity for spatializing time has gotten them into. For Bergson, the governing notion is pure duration, so conceived as to be, as it were, the foil of space.

I. Hegel: Time as the Truth of Space

There is a notion that time *is* process, physical eventuation—be it realized in the locomotion of a single mobile or in the directed succession of states, as in the dispersion of heat and the expansive motion of matter in the universe or in the development of organisms with respect either to individual ontogeny or collective phylogeny. It arose under the first heading of this chapter and will return in its last section, on Bergson. It is the substantive complement of the reductive notion that natural time is a clock or its readable positions. The critique of the latter operating assumption consists simply of pointing out the undenied fact that it is a mere circumvention of the question "What is time?"

The idea of process-time, on the other hand, is an answer to that question. It is intended to impart to changing nature at least one feature that will be found in human time: a forward-directedness in which perhaps some memory-analogue can be discerned. The critique of process-time is more tricky, because one must clue out what it would mean to deny that time is process and its converse. The thought comes to aid that the key is in the degree of abstraction. If time is taken concretely—as Whitehead sometimes takes it—so that the physical bodies or their states themselves in their alteration are said to be time, the criticism amounts to the observation that nothing is gained by tagging them as time (Whitehead 1927, 35). The claim is then on the level of the identification of time with the clock and its positions in being a way of masking the problem. But if the meaning is that there is some separable element, derived by abstraction from the physical process, that is peculiarly responsible for its progress, then something has been accomplished.

Hegel, and to my knowledge only Hegel, does just that. He works out the conceptual genesis of pure external time, of a becoming that is in external nature, and yet is not the becoming of something, that is, the material process.

The genesis of time is set out in the second part of Hegel's system, in Part Two of the *Encyclopaedia of the Philosophical Sciences* (1830), the *Philosophy of Nature* (paras. 257–59; the "Additions" sometimes quoted are not directly by Hegel, but were compiled by Michelet for his 1847 edition). No other philosophical writing on time equals it in difficulty. Heidegger, who thinks it is in part unintelligible, explicates these paragraphs in *Being and Time* (82; also in *Logik*, 20), not sympathetically but penetratingly. Derrida, commenting on Heidegger's commentary, is also helpful (Derrida 1968, particularly 25–46). I shall do my best to give an intelligible, clear rendition of Hegel's three paragraphs.

A setting is, however, needed, since every dialectical moment carries within it its "before and after" in the system. For the dialectical approach locates the moving principle *within* each concept, and to think dialectically is not to order thoughts externally by analysis or composition, but to watch them develop themselves, reanimated, so to speak, by the conceptual receptivity of the thinker. Hegel's name for the thought-moment capable of dialectical motion is "Concept," and the developmental motor of a concept is "negation," movement by self-exceeding, which makes the concept more concretely determinate—more thought out, so to speak. For no sooner does a moment of thought have its position in thought, no sooner is it posited as a determinate, definite concept, than it is seen that this positing occurred through a negation, a delimitation. Thus the *Logic*, which as Part One of the *Encyclopaedia* precedes Nature, begins with Being, the least determinate concept of all, in which all definite thought is only implicit, that is, as yet undeveloped. Therefore, no sooner is Being posited than it is revealed both in itself and to us as being as yet absolutely negative, and so Nothing. And Nothing is immediately revealed as being just what Being was and therefore forming a unity with it. But the unity of Being and Nothing is Becoming, which is "the first concrete thought," the first concept having internal distinctions within itself (86–88). Becoming is that reciprocal passage from Being to Nothing and Nothing to Being that we can recognize as the most abstract, formal pre-formation of change: To change is, minimally, both to be and not to be (something), undividedly and continually. That is, of course, why this reference to the *Logic* was necessary.

Logic is atemporal, and time does not turn up in Hegel's *Logic*. It turns up in Nature. Nature itself is developed dialectically out of the final and fully determinate Concept reached in the dialectical development of the Logic, which Hegel calls the Idea. It is not part of this exposition to dwell on the question whether this development is really dialectically smooth or is in fact a special leap. Suffice it to quote Hegel's statement in the *Philosophy of Nature*:

> Nature has presented itself as the Idea in the form of *otherness*. Since therefore the Idea is the negative of itself, or is *external to itself*, Nature is not merely external in

> relation to this Idea (and to its subjective existence Spirit); the truth is rather that *externality* constitutes the specific character in which Nature, as Nature, exists. (247)

We are to understand (1) that Nature is, as dialectic requires, the negation of the Idea; (2) that this particular kind of negation is otherness, which Hegel understands as externality—the Idea has gone outside of itself, beyond the confines of pure thought; and (3) that the kind of negation the Idea has undergone determines the character of Nature not only extrinsically, as the sphere that is the other of the Idea, but also intrinsically, as the realm of the *self*-external. I repeat here that for present purposes the possible critique of these dialectical transitions is suppressed; what matters is to see where Hegel is going.

And that is toward conceiving space as the first (and, indeed, also the second) concept in Nature, as parallel to logical Being (and, because of its indeterminateness, to Nothing as well) in its primal position in the development and in the total implicitness of all its features. Space, or external Being, has but the one feature of Nature, its self-externality or side-by-sideness (254). This Space is readily recognizable, for most people who think about space, including Bergson, characterize it as that all of whose parts are outside each other.

Thus time is consigned to second place in nature (though not irredeemably so in the whole system). Time comes out of, or after, space, but it also comes before motion, as motion will come before matter. In parallel with the *Logic*, time is Becoming, external Becoming. It is thus precisely that *pure abstract motion of physical change* that those who identify time and process ought to be looking for if they want to make their identification more than a renaming of the directed variability encountered in the world.

It remains to set out as intelligibly and briefly as I can how time comes *out of* space. There are two strands to follow: (a) the characterization of time as Becoming, and (b) the actual springing of time from space.

(a) Time is directly "intuited becoming" (*angeschautes Werden*, 258), that being that in *being* is *not* and in *not* being *is*. The fact that time is *intuited*, "looked at," distinguishes it from the Becoming of the *Logic*, which was *conceived* as a unity of Being and Nothing. This new Becoming, which arises out of space, is intuitable because it is external, and so it is the other of thought. It is again an almost undifferentiated, primitive abstraction, standing near the beginning in Nature of that process of dialectical differentiation that logical Becoming had to undergo in thought, but it is now an "*actually existent abstraction*" (258, Remark); it is beyond mere thought—outside.

In arising from space—just how will be shown in (b) below—time loses the thought-unity of logical Becoming. For its externality differs from the indifferent, side-by-side, all-at-once, static difference of space; intuited Becoming is determined by *momentary* differences, differences that "immediately cancel (*aufheben*) themselves as external, however as self-external." Time, as intuited mere Becoming, is "pure being-within-self as sheer coming-out-of-self." Hegel is giving a dialectical description of the succession of time as distinct from the simultaneity of space. Space is being animated, or, better, is animating itself: "Difference has stepped out of space. This means that it has ceased to be this indifference; it is for itself in all its unrest, is no longer paralyzed" (257, Addition). Time is being described as a kind of self-motor that projects itself from moment to moment, always restlessly different and yet always remaining the same.

As a first, abstract or "ideal" intuition, time is also described as "the nonsensuous sensuous," a "*pure form* of *sense* or of *intuiting*" (*des Anschauens,* 258, Remark). The half-way attentive reader of a translation of the *Philosophy of Nature* will be brought up short here. Surely intuition is being used in two senses: as the object being seen and as the activity of looking. And indeed the German is used passively and actively within the same five lines: Time is both the "intuited being" (258) and the form of "intuiting" (Remark). In fact, Hegel changes the Kantian "form of intuition," a receptive mode for taking in sensory stimulation, to make it unambiguously active. Is time an object or a subject? Strangely enough, it is both. Insofar as *we* are recapitulating the genesis of time in thought, it is certainly an object, one to be thought of *by us* as intuitable. But regarded for itself, "the distinction of objectivity and a subjective consciousness contrasting it does not apply to time." Space might be thought of as "abstract objectivity," but then time should be thought of as "abstract subjectivity" (258, Remark). The meaning is that space is without consciousness and only an object of (someone's) intuiting, but time is beginning to be subjectlike. For subjectivity, as opposed to the positive "thisness" of objectivity, is for Hegel marked by just such a self-negating unrest as defines time.

To show that Hegel means what he says—however that might become intelligible—let us turn to his treatment of the phases of time, for they are generally agreed to be that aspect of time most inseparably related to consciousness. Hegel ingeniously explains past, present, and future by pulling apart the logical unity of Becoming into the continual exchange of coming-into-being and passing-away. When Being is regarded as the starting point and Nonbeing added to it, we get the past of natural events and of history—actuality with a negative sign. When the position is reversed, we get the future—what is not that is to be. The present is the middle term, the indifferent unity of both; it *is* only because the past is not, and simultaneously, it *is not*, and this nonbeing of its being is the future. This present or Now is therefore a "negative unity" (259, Addition). Thus Hegel seats the phases in external Becoming. And that motoric negation is testimony that this Becoming is a sort of protoconsciousness. Later, in the realm of mind, it will impart to the phases the persistence they gain through remembrance and hope, which arise with the development of subjective imagination (259, Remark). Hegel sums up this subjectlike time in the most elementary formula of self-consciousness: "Time is the same principle as the I = I of pure self-consciousness, but . . . still in its uttermost externality and abstraction . . ." (258,

Remark). That is to say, space-emerged time is *self*-differentiation, albeit monotonously reiterated. In dialectically extruding time, space is beginning to come to, is sending forth signs of consciousness and eventual self-consciousness (p. 206).

(b) How does this miracle of the incipient return of Nature to thought as Spirit come about?

> The truth of space is time, and thus space becomes time; the transition to time is not made subjectively by us, but made by space itself. (257, Addition)

I quote this passage to show that Hegel actually says these things, and as he says them, so they have to be understood. In Hegelian language, a dialectical stage is "the truth of" the succeeding stage if it makes good a deficiency in the former. The defect of space was not that it was not infected with negativity. (Recall that negativity, the timeless motor of thought and a moment of becoming, is determination by self-exceeding and delimitation.) It was rather that here negativity was dead, featureless indifference, indifference to each other of the parts of space as well as of the three dimensions (254–55); space does not succeed in getting beyond its own side-by-sideness.

Time makes good this defect. It is "what is negative in itself, . . . negation of negation, . . . self-relating negation. Negation in space is negation attached to an Other; the negative in space does not therefore receive its due" (257, Addition).

We are to understand that in space the negative is not intensive enough to overcome the "paralysis" of spatial indifference. Now the proper negation of space is the point, but where the point remains related to space even as it negates space, it merely turns space into a manifold of points (254, Addition) or, as the head of a line, draws itself out into the dimensions of space, both without leaving it (256). But when the point is regarded as self-external, as self-canceling, it leaves the inert side-by-sideness of space and becomes time. It is here that Hegel says that this transition is not made "subjectively by us but made by space itself" (257, Remark), and that "time is the same principle as the I = I of pure self-consciousness" (258, Remark).

There are two ways to come to grips with this exposition of the genesis of time *and* subjectivity. One is purely formal. In asserting its self-externality in opposition to that of indifferent space, the point becomes a principle of difference, of unrest, through the perpetual intensivity of its drive: It continually negates itself and asserts itself into a medium that is no longer that of spatial indifference; it becomes *eventlike*. And it can be thought of as doing this itself, because negation of negation, double-negativity, self-negation, are terms descriptive of the dialectical motion, however rudimentary, of Spirit, as Heidegger demonstrates (82 b). It is this Concept thinking itself, this self-consciousness, whose most abstract form is expressed in the formula I = I, a formula combining self-differentiation with self-assimilation. And the point exhibits the formally identical impulse to self-determination that constitutes selfhood.

But of course, these merely formal identifications, though a prerequisite to the argument, are not very satisfactory. How can we understand the point-origin of time and its characterization as intuiting-intuited becoming in those accessible figurative terms that Hegel often decries and so often employs? There is a way.

In its self-negation the point, *a* point, differentiates itself from all the ones lying dead in space. It is now here and now there (Heidegger, 82 a). It rises out of its space-bound position and passes through space: "Time is *spacing*" (Derrida 1968, 43). We *think* through the nature of space and come on something like primal passage: now here, now there. And we *imagine* it as pure Becoming: intuited Becoming. But we could not think through to the truth of space did space not have this negation, this thought-determination within itself. As a concept, albeit an external one, space may well be spoken of as going beyond itself through its own dialectic. Thus out of space develops time, as a passage through space from beyond space: the form of intuiting, which has the rudiments of attention to an other—to space—that is a mark of subjectivity, together with the negativity, the unrest that is a mark of selfhood. From its first genesis time, pure Becoming, behaves like incipient Spirit. We are watching our own dialectical origin.

The next stage, which we do not need to follow, includes the reunion of space and time in the Here *and* Now called Place. Then follows motion and the transition from ideality to reality that comes with the introduction of matter (260 ff.). It is worth pointing out that Hegel's genesis of the physical world follows the history of physics: first kinematics (Galileo), then dynamics (Newton).

This is the moment to say that this project of giving the conceptual genesis, or in his term, the "dialectical" development, that will put time *into* nature, is not, finally, persuasive to me. For it only works within a system that conceives of nature as implicitly subjective, as being determined by what is non-natural—by Concept and Idea—in such a way as to bear within itself a striving to come, or rather to return, to consciousness and self-consciousness. This notion is different from the trust that nature is amenable to intellect. For it means that nature is not only intelligible but is incipiently egolike, that is, subjective. But nothing is better designed to drive one into a sense of the opposition between nature and humanity than an immersion in the question of time. Hence Hegel's development of time is indispensable to following out the thought-connections that *might* establish time in nature, but it is not sufficient for that complete reflection that *must* do so.

A brief last word about an inevitable question: Is this first and abstract occurrence of time in the System of the *Encyclopaedia* a recognizable dialectical predecessor of the grander version that appears in that prelude to the Hegelian System, the *Phenomenology of the Spirit* (1807)? The answer has already been prepared: Time is always what it was from the first—formally speaking negation of negation and dialectically speaking empty intuited Becoming. It begins when the Concept or Idea goes out into Nature; it ends when the Concept has recollected itself into itself without remainder. All this is clearly said in the last chapter of the *Phenomenology*:

> Time is the Concept itself that *is there* and which presents itself to consciousness as empty intuition. For this reason Spirit necessarily appears in Time, and it appears in Time just as long as it has not *grasped* its pure Concept, that is, has not annulled Time. (para. 801)

Midway along the dialectical road from time's first leap *out of* space to the end of Time, Spirit has come on the scene and into Space and Time. It is now the Idea returned to itself from Nature, and, enriched with all the natural determinations, it is on the way to self-consciousness and all its recollections—a *being* that is *there* (*Dasein*) to act in the world. Thus the remote and proximate presuppositions of Spirit are the Idea and Nature respectively, and its development is set out in the third part of the *Encyclopaedia*, the *Philosophy of Spirit*. Time appears briefly at the beginning of the major phases of Spirit in the form belonging to each context. So the memory is treated in the *Philosophy of Spirit* early under "The Theoretical Spirit" (452). But time goes quickly underground to work invisibly away as the formal motor of change until it ends in the fulfillment of complete self-knowledge.

Yet one more question arises: Is time as intuited Becoming powerful or not? Hegel answers clearly—in the paradoxical Hegelian mode: "though time is called omnipotent it is also completely impotent" (258, Addition). It is impotent because it is not one entity, not a flowing stream in which things are placed to be swept away. They perish because they themselves are temporal and finite. By this announcement those who assign an independent moving power to time prove themselves false followers of Hegel: "It is therefore the process of actual things themselves which makes time. . . ."

And yet, there is a sense in which this Becoming that is found in all that is real is a power that dominates reality:

> What is real is limited and the other to this negation is *outside* it; therefore the determinateness in it is self-external and is consequently the contradiction of its being; the abstraction of this externality and unrest of its contradiction is time itself. (258, Remark)

In other words, time is the dialectical incompleteness of real being that pulls it from beyond its limits into the assimilation of all its successive contradictions. Its power is the power of negativity: *logically directed and sensorily intuited Becoming*. When Nature has completely reabsorbed the logical Idea and has become the fully determined Concept, Time ends and Eternity supervenes.

Bergson and Hegel are poles apart except in their interest in the relation of time to space. Their antithetical use of the word "duration" is a sign of their difference—and of a certain connection. For Hegel duration is time in which there is little motion; the worst and the best things endure: Nature in its mere repetitiveness and the universal Concept in its life. Absolute duration, self-subsistent and reflected into itself, is Eternity. "Reflected into itself" means "self-related." Eternity collects the phases of time back into itself, into a timeless Now (259, Addition). For Bergson duration will be the Now of human life insofar as all past moments are collected in it through a reiteration of interpenetrations. It is a passing eternity, so to speak. Their view of the spatialization of time is also curiously similar; Hegel denies the possibility of a science of time because such a science must express time through configurations; this can only be done if the understanding (the nondialectical reason) paralyzes time and reduces its negativity to a unit, which is an inert One, "the utmost externality of thought," the sort of bare item that can be externally combined as in arithmetic. Bergson will inveigh against just such an externalizing configuration of time in indifferent space, but from a quite undialectical point of departure.

II. Bergson: Space as the Falsehood of Time

Bergson's mission is the suppression of extensive space in favor of intensive time. The relevant text is his essay *Time and Free Will* of 1889 (ch. II and Conclusion). In this work he carries out his mission by mingling introspective observations (psychology) and a conceptual analysis (philosophy) with an exhortation to recover the fundamental self (rhetoric).

Extensity is the essential characteristic of space. It betokens that the parts are outside of each other and that the whole is the sum of its parts. Intensity is a mode of being in which the parts interpenetrate each other and the whole is that concretion of parts. These terms used to be more current in science than they are now. An example from physics of an extensive magnitude is the spatial dimension length, measured in linearly additive units; hardness, on the other hand, may be described as an intensive magnitude, since its units of measurement, degrees, belong equally to all parts of a homogeneously hard body, and the degree of hardness of the whole cannot be increased by adding the degrees of the parts. Bergson frequently uses illustrative metaphors from the natural sciences, but his notion of time as intensity is, of course, not exhausted by the physical example; strictly speaking, he says, "pure duration" (his name for intensive time) is not a magnitude at all (106), not a quantity but a quality. The moments of time are neither a numerical multiplicity nor a continuous line. Bergson is, however, fully aware that we symbolize time to ourselves quantitatively. The convenient abstractions of ordinary life and of philosophical discussion, the repeated associations of ideas that logic and language call for, and the externalization of inner states that are the conditions of social life—all these tend toward forcing us to image our conscious states in a homogeneous medium. That medium is space.

Thus we exist on two levels. One is the level of deep-seated psychic phenomena; here conscious states are qualitative, directly perceived, fluidly interpenetrative, and personal—felt as true time, as real duration. The other level is that of the psychic surface; here conscious states are quantitative, indirectly perceived, solidified in mutual externality, and impersonal—represented "by a refraction through space" (137) as undynamic simultaneity.

"There is a real duration, the heterogeneous moments of which penetrate each other" (110). Space, on the other hand, is homogeneous. The time of the ego and the space of objects are givens for Bergson, given respectively as psychically internal and external. He does not derive them dialectically, as Hegel derives time from space, nor does he ground them in some sort of Being—that is what makes his treatment a philosophical psychology rather than an ontology of time. In fact, he evinces a revulsion from analyses and classifying language; for example, he describes vividly how the mutually permeating elements of a deep feeling lose their life and color when enumerated and named. Nonetheless, he does analyze closely how time comes to be spatialized, how one of these radically different entities comes to be represented by the other. The problem is "the possibility of setting out in space, under the form of numerical multiplicity, what we have called a qualitative multiplicity, and of regarding the one as the equivalent of the other" (124).

He uses the physical process of "endosmosis" as an analogy, one that is, however, not to be taken as really explanatory (109, 228). Endosmosis is the flow of a fluid through a permeable membrane toward the side of a fluid of greater concentration with which it mixes. We set time out in space, and so produce the mistaken idea of a homogeneous inner duration, by allowing duration insofar as it is succession to flow into space insofar as it is homogeneity. Thus we get the mixed and contradictory idea of a measurable time, of succession in simultaneity. Simultaneity, the system of spatial positions at any moment of consciousness, is the external present, the only "duration" existing outside us. It is the intersection of time and space, so to speak. Space itself and its contents know no successional duration; external things do not endure in the living sense but they do present a changing face to us (227). Note that external Becoming is here acknowledged as the mystery that always remains when time is internalized, a mystery that must come to the fore in Part Two (p. 201).

Bergson analyzes how the injection of succession into simultaneity in fact comes about. Thus his chapter on time begins with a long consideration of number, for number is the point of departure for the self in its fall into space (p. 99). We may think that numbering is a successive, nonspatial activity, but that cannot be so. A number is a multiplicity of identical units, and we can form no picture of such a structure without the intuition of space in which the units are juxtaposed. For in the succession of pure duration we have in mind only one unit at a time; to make the multiplicity reach the final number we must deposit each counted unit in space, either as a real object in real space or as a mental image in ideal space, or even, to get abstract numbers rather than concrete collections, just as points in either space. Successive moments can indeed be conceived independently of such spatial registration, but they cannot add up to their sum, for each moment disappears forever into the past unless it leaves its spatial trace in the present. One might put Bergson's notion this way: Space as a simultaneity is the external memory for additive multiplicities. It seems to me that the notion is not entirely lucid: If a spatial line-up is necessary because each moment of a succession vanishes, when exactly do we actually apprehend the additive configuration? Surely not in one indivisible present moment—but by the previous argument not before that either.

In any case, we then go on to make the mistake of thinking that our counting was really done in time, when it was in fact done in space. Furthermore, as we deposit the connected point-units in space, they are indeed discrete, as the units of a number-quantity ought to be. But they "have a tendency to develop into lines in proportion as our attention is diverted from them" (83, p. 209), that is to say, they tend to connect themselves with each other so as to produce a line-continuum. Thus, when the count is complete the number has not only a spatial configuration but a continuous structure. This theory—a Pythagorean revival, for the ancient arithmeticians thought of numbers as laid out in space—is as different from Hegel's as possible. For Hegel pure number is prior to and independent of space, and is in fact developed within the logical, not the natural part of the *Encyclopaedia* (102), though as set out above, arithmetically operable number is spatial (p. 29). Bergson might, of course, say that it is the psychology of counting, not the genesis of numeration, he is explaining. But if Hegel is right that number is first a logical concept, then a prespatial understanding of number is possible, and Bergson's exposition is incomplete.

In any case, once again we then go on to transfer, illegitimately, the spatialization and the continuous representation of number to our affective psychic states, that is, to duration. We count these states by a process of symbolization. Thus sensations, for instance the strokes of a bell, are (usually) ranged by us in an ideal space where they symbolize succession for us, so that we fancy we are counting pure duration. Thence arises our notion of a homogeneous, measurable time—an impure duration, one might say. It borrows from space its essential property, that of being an empty, homogeneous medium for differentiating identical simultaneous sensations. In truth, however, affective states are neither countable nor linearly representable; they are rather a total "qualitative impression produced by the whole series" (86).

Thus the time of natural science arises as an illusion, for in space there is no past, only the present position of the clock hands. The "living symbol" (110) of this homogeneous pseudoduration is motion, which is real only for the spectator, who, by a mental synthesis, psychic and unextended, extracts from external motion its "mobility" (111), its intensity as a quality. Motion is merely space transversed. Science has to eliminate pure duration to make its time homogeneous, and Bergson points out that physicists in fact do just that. They leave time aside and deal with its measurement in terms of a noting of positions coincident with a psychic state, in short, in terms of an observer reading a clock at a signal, much as described earlier in this chapter.

Bergson is, as was said above, perfectly candid about the fact that his internalization of the quality of motion leaves external change as a mystery. All we know is that there is in external things "some inexpressible reason in virtue of which we cannot examine them at successive moments of our own duration without observing that they have changed" (227). This candor is worth a great deal, for it makes explicit a problem faced by anyone who places time in the soul.

Bergson treats internal or "pure" memory in *Matter and Memory* of 1896 (ch. II). Memory is a "privileged problem" since it lends to present perception its subjective character while differing from it radically. Perception is the present contact with the present object. It is a duration spanning the perception of the immediate past and the determination of the immediate future—what will appear later in this book as the "specious present" and will be subjected to deep analysis by Husserl (p. 142). Perception is always impregnated with interpretative and fulfilling memory-images. These memory-images themselves "materialize" pure memory, which is in itself latent and powerless. It shows itself in the nascent image and eventually in bodily action only when it can become useful to the present. Pure memory is thus past that has ceased to be presently useful. It is not lodged in the brain but in the unconscious, an unconscious that differs from Freud's powerfully interfering inferno in being dead until revived by present need.

Again, Husserl will deal with just such issues, the durational present and the relation of memory to perception, but more penetratingly. One more partial similarity bears mentioning, however. Bergson draws a diagram in which an axis representing memory is erected at right angles upon a base representing objects simultaneously in space (184, fig. 3). The figure represents the double line of states and things that intersect in perception, in actual present consciousness. Husserl's diagram of time will label the axes memory and time, and thereby hangs a tale (p. 135).

I should note here that, whereas Hegel's dialectical development of time has found few takers and the *Philosophy of Nature* is not well regarded by scientists, Bergson is frequently referred to by a whole line of philosophers thinking about science, and recently even by scientists. Chief of the philosophical writers are Samuel Alexander (*Space, Time, and Deity*, 1920), Alfred North Whitehead (*Process and Reality*, 1929) and George Herbert Mead (*The Philosophy of the Present*, 1930). What brings them together is an interest in living events and processes, in the relation of time to space, and in the human observer of nature. More recently the physicists David Bohm and Ilya Prigogine have been interested in a turn of physics toward process-philosophy. To all of these Bergson is the avatar (Griffin, intro., 15 ff.).

Bergson's central notion is that of pure duration, of a time that has successive moments that interpenetrate each other. To me this notion seems muddy, though the motive is clear. It is the salvation of the deep, fundamental human self from a fall and dispersion into a second, spatial and societal self. When we become external to ourselves we become passive rather than active; we lose our freedom. It is a question whether Bergson's salvation is in fact salutary. But in spite of doubts one might have about the human consequences of retracting inward from space, and in spite, too, of the disproportion between his insights and his grounding analyses, there is no better writer to lead one from external time to that humanly constituted time that will be the subject of the next three chapters. In particular, his trenchant criticism of external time, his acknowledgment of the mystery of exterior change, and, above all, his insistence that interior time cannot be simply extensive—these are good notional equipment to bring to the study of inner time.

Chapter Two

Aristotle and Kant: The Counting Soul

Here again are two philosophers whose understanding of time touches at this crucial point: The natural world is in time, and that is the doing of the human being, whose ability, as soul or as subject, to sense magnitude and to count in numbers is at the root of its cognitive capacity.

Again I am far from arguing influence, the most tiresome and unsatisfying of explanations—that in terms of purloined or borrowed ideas. To be sure, little can be said in philosophy that is not indebted to Aristotle; and so, as Whitehead said that all philosophy is a footnote to Plato, Immanuel Kant acknowledged Aristotle as a direct predecessor by saying that he was the first thinker for whom "philosophy [was] work," that is, a methodical pursuit; moreover his results, particularly his establishment of the categories of reason, are still usable, though he is wrong in extending them from the sensible field in which they are valid to the supersensible realm for which they are inadequate ("Of a Newly Arisen Refined Tone in Philosophy," 1796). Evidently Kant regards Aristotle (and Plato even more) as exceeding the warrant of thought and therefore as preparing the problems that he intends to solve. And perhaps that is what my claim to similarity really amounts to: Aristotle poses implicitly certain definable problems that Kant squarely confronts. But that is no small commonality.

Thus what comes to the fore in Kant's *Critique of Pure Reason* is the solution to a specific difficulty only implied in Aristotle's treatment of time, that of explaining how the soul counts and, in particular, how it counts what it senses as continuous magnitude. My supposition is that the unresolved questions a philosopher sets up, the well-delineated abysses he opens for future delving, are equally with his explicit teaching a part of the burden of his thought. They are the ghosts of philosophies to come—and therefore I think that the opening of such depths by a philosopher is quite as interesting as are his fixated and plausible teachings.

Nonetheless, even if Aristotle's account of time is taken in itself and as a fairly complete whole, it is nothing short of spectacular. For the sort of deconstruction by sober analysis to which he subjects time has radical consequences. The Loeb translator says with charming cluelessness: "As to time Aristotle enters into no profound metaphysical speculation as to its essential nature" (I 378). We will see that therein lies the point and the depth of Aristotle on time.

The surface similarity between Aristotle's and Kant's treatment of time will be, to anticipate the thesis to be pursued in this chapter, the relation of the apprehension of time to counting, while the deep similarity will concern the constitution of human knowledge and ultimately of human Being. My account for Aristotle will naturally depend primarily (though not exclusively) on chapters X–XIV of the Fourth Book of his *Physics*, in which time is the theme. For Kant it will depend on the time section of the Transcendental Aesthetic, on the Transcendental Deduction, on the Schematism of the Understanding, and on the Analogies of Experience of the *Critique of Pure Reason*. Once again I will try to give a balanced account of the two theories while bringing out the elements that seem to me both fundamental *and* common.

A. ARISTOTLE

I. The Being of Time

Aristotle begins his inquiry into time by "taking up the perplexity, looking both to time itself and to the 'outside arguments'" (*exoterikon logon,* 217 b); the perplexity is whether time is among the things that *are* or those that *are not.* The question concerning its nature is to come afterwards. Few writers begin their inquiry so radically.

Let me state up front: this "nature" will be a non-nature, and only its explication will give the definitive answer to Aristotle's first perplexity: As a non-nature, time will be the number of motion; motion, in turn, is a continuous magnitude displaying "before and after"; its number, in turn, will come about through the soul's counting. This understanding, which is a definition only in the very largest sense of being a collection of notions exhaustively circumscribing the term, will put so many barriers between time and Being that the answer to the question, Is time a being or does it have Being of some sort?, got by clueing out the intermediate terms of magnitude, motion, and number, can only be: No, except in the most equivocal and remote way. I should say here that I avoid the word "existence" with respect to time in this part of the exposition because Aristotle will use the Greek word "exist," in a different signification, not of time but of motion (221 b 3), and because in one restricted philosophical use it means "being here and *now*," "being in space and *time*" (p. 176). It is therefore useless for asking the question concerning the Being of time. "Being" (*on*) is Aristotle's word for what we vaguely catch by "existence" in the widest sense, which includes "actuality."

Whose are the "exoteric arguments" against the Being of time, and what's wrong with them? Thomas Aquinas takes these arguments to be not popular but "sophistical." I imagine that he is thinking of the same sophists who turn up again later on in the book with bad arguments (219 b 21; *Commentary on Aristotle's Physics*, para. 558). Their points certainly don't seem much like folk wisdom about time. Judging from the Greek poets' references to time as well as from our own current speech, the popular view of time sees it as intensively potent and even personal, and is not at all inclined to evaporate it (p. 193). That is why Aristotle's deconstruction of time seems so, one might almost say, scandalous.

These sophistical arguments say things like this: Time does not exist (here, for once the right word: "has no presence") because some of it has gone and the rest has not yet come. Moreover, Now isn't a part of time at all, for Nows don't add up. That is to say, the sophists fix on the fact that the Now is not what mathematicians call a Eudoxan quantity, one which can be added to itself to become as great as you like.

These points seem to be pretty good; Aristotle himself says in the *Categories* that no part of time stays put, and so it has no "position" (5 a 28). I think that Aristotle dismisses the sophists—he certainly doesn't refute them—because their proponents aren't *radical enough.* They look to a sort of word-play on the tenses of "to be," and depend on the fact that "was" and "will be" *is* not "is." Hence the argument is perilously close to being a paralogism, for it relies on time being something in the past and in the future in order to refute its being in the present. Those of the "outside arguments" that involve legitimate perplexities about the Now, whether it is always the same or always different, will be thoroughly analyzed by Aristotle. But their contention that time is "scarcely" or "obscurely" just isn't sufficient. The sophists' accounts are feeble, not false.

The more positive theories that Aristotle mentions are, in his opinion, inadequate as well and leave the nature of time "unclear" (218 a 34). For example, Plato and his Pythagorean predecessors identify time with the heavenly revolutions or even with the heavens themselves. Aristotle thinks this is naive (218 b 8). Why naive? Probably because it is *too* radical: It reduces time to a heavenly clock, as I've tried to show in the previous chapter. So we have, by way of earlier accounts, some too feeble and others too forceful.

For us to follow Aristotle in asking about the Being of time and to be entitled to say that something is a nonbeing, a nonentity, we must recall what Aristotle means by a being or an entity. As he says right in the *Physics* (185 a 22), and often in the *Metaphysics*, "Being is said in many ways." It can mean practically anything: what is only on the way to being—motion is of that sort—and even what is altogether not (*Metaphysics* 1003 b 10). But Being does have a primary meaning: It is substance, that ultimate independent individual entity that is no longer predicated of something else (1028 a).

Now, as I have said, time for Aristotle is certainly not that sort of a being—the sort that can be met with in myths and poetry and in ordinary figurative speech, such as Father Time, or even an invisible force or a fluid substance, such as Newton's equable flux. Aristotle does allow that time may be thought of as having a power of decay (*Physics* 221 b 1, 222 b 19 ff.), a power remotely analogous to our physicists' "arrow of time," entropy in the loose sense, meaning nature's direction toward disorder (p. 15). So it might, Aristotle concedes, rather than being called, as some do call it, "the wisest thing," be called, with Paron the Pythagorean, the most ignorant, since through it everything goes downhill. For all coming into being needs an articulable cause, while decay does not seem to require an explanation: Motion all by itself "makes to stand out" of its place what is at hand (*existesi*, 221 b 3; here is Aristotle's use of the word "exist" referred to above). "All change is by nature out-standing" (*ekstatikon*, 222 b 18); these are attempts to render in English Aristotle's terms for the deranging or dislodging effect of time; the two *sta* words will have a vivid future in time-inquiries (p. 104). But even in its deteriorating respect time does nothing in itself, Aristotle observes, for this effect is merely incidental to motion (p. 53).

That time *belongs* to change is the beginning of Aristotle's own inquiry—and also that it *is* not change. For change takes place only in particular mobiles, in the things or beings that can move or change, while time is everywhere and in all things alike (218 b 9–14); moreover change can go faster and slower, but not time.

Just what is the relational mode in which time stands to motion? No term for it, such as affection, accident, or property, is named. Since, however, time will turn out to be a quantity and is so named in the *Categories* (5 a 6), it can be, whatever the mode of attribution may be, "said of" motion—though motion is itself, at least in one aspect, a quantity. Thus time is some sort of secondary property, one not to be directly sensed as motion is. This relation of time to sensation will be explicated below (p. 45).

Since time is—however that may be—closely involved with motion, its analysis rightly takes place in the *Physics*, whose subjects are nature, or better, natures, that is, all beings that have the source of their motion within themselves, and the motions that occur in the world. It is not unsignificant that the treatment of time comes immediately next in line after that of place and void. For motion occurs in place (*topos*), and place raises the question of Being or Nonbeing, namely with respect to emptiness (213 a 11). Void is shown to have no "separable" being (216 b 21), and the inquiry into time, which "takes up from" the demonstration "how [void] is and how it is not" (217 b 27–30), seems to be somewhat in the same mode: It too begins as an inquiry into Being and Nonbeing. Aristotle says the same thing at the end of his study both of time and of place: "It has been said that it is and what it is" (213 a 11, 222 b 28); moreover, he says of place as well that, although it appears to have a certain power (208 b 11), it too is full of perplexity concerning its Being and "hard to grasp." This argument, that the positioning of the inquiry into time shows it to be among the things that *are* only dubiously, is not very strong. But it figures somewhat in the demonstration of the claim that for Aristotle time is not even remotely substantial.

What then is time, such as it is? Let me recapitulate briefly Aristotle's answer and then go on to consider its elements. For the implications of the terms bring out the meaning of his answer.

1. Time cannot be motion itself, though it belongs to motion. Nor, again, can time ever come about in the absence of change; this second condition turns out to raise a deep problem (p. 46).
2. Then the question becomes: What belongs to motion that makes time possible? Motion is a continuous magnitude (p. 40) because it happens primarily in place, that is to say, it occurs from a here to a there that between them mark a continuous and measurable path. Moreover place is what primarily has "before and after," the successive positions that display the ever-otherness of change. Therefore the occurrence of motion implies place, and it is place that supplies continuous path-magnitudes as well as directed successions of before and after—not the only such successions, but the primary ones.
3. When we mark a motion by marking its limits "we take note of time" (*gnorizomen ton chronon*, 219 a 22). In other words, when we "take in" or know (*noesomen*, a 27) these boundaries by Nows that are one before and the other after, we speak of time; it is what is measured between them. And when we count the Nows we get a number. That number *is* time. Here, once again, arises a deep problem (p. 45).
4. So, collecting all the terms, time is that by which motion has number in respect to directed succession of "before and after." The number is gotten as a count of the measures, (221 b 8) marked by the soul using the Now as marker (219 a 23 ff.; the term used over and over for marking is *horizein*, from which comes our "horizon"; it means "to mark out a boundary or a limit").

II. The Elements of Time

What could be more plausible, even flat? For the depth of it is out of sight until we take up the elements one by one: (1) magnitude, (2) motion, (3) before-and-after, (4) number, (5) now, (6) soul, and (7) clocks. The sixth of these is not a constituent of time but its producer, and therein lies the chief problem to be pursued.

1. *Magnitude*

Magnitude is the supportive tissue of the sensed world. Every sensible thing is a magnitude (*On Sense and Sensibles* 445 b 9), and every magnitude is, at least in principle, sensible (448 b 14). Thus being a magnitude and being sensible are convertible.

Magnitude is one of the common sensibles—as is motion, which, however, "follows on magnitude" (*Physics* 219 a 12). Common sensibles are not sensed directly by their proper sense, as color is by the eye, but by the "common sense" (*On the Soul* 425 a 14 ff.). It is a root sense; but although it is a separate sense, it is a common concomitant of, or is incidental to, the other senses' sensing as well as of motion; ultimately all sensing is through motion. The common sense is, moreover, the sense that discriminates all sensations, and it is the power of awareness, of sensing that we sense. Its discussion in *On the Soul*, foreshortened though it is, might be thought of as analogous to Kant's Transcendental Aesthetic, his analysis of the roots of sensation in the first *Critique* (p. 61).

The chief property of magnitude itself is that it is continuous (*syneches,* literally, "what holds together," *Physics* 227 a 10 ff.). It is one of the two branches of quantity (*poson,* "how much," *Metaphysics* 1020 a 7 ff.). A quantity is that which is divisible into its constituent parts. If these are ultimately discrete and indivisible, they are units and the quantity made of them is a number. If the parts are indefinitely divisible the quantity is continuous and is called a magnitude (*megethos*).

The prime example of magnitude is a geometrical line. Any part of it can be further divided. Its boundaries, points, are not parts of the line and are not homogeneous with it in dimension. Any parting of the line produces one boundary, one point, for both parts. This point, the crucial element in the time-analogy to come, plays a double role: It divides the line, and it also holds it together. It will be the model for the Now, as the line will be the model of motion in the later books of the *Physics.* Continuity is what makes magnitude the supporting fabric of the world, for it is the expression of the world's total plenitude. Its character will descend to time in this chain: continuous magnitude → path → motion → time (*Metaphysics* 1020 a 32). "Next in succession" (*ephexis*) is, from the point of view of thought, prior to continuity and belongs to a more primary kind of being such as number, whose elements do not touch or coalesce (227 a 18 ff.). In other words, discreteness is closer to thinking than continuity: "For thought (*noesis*) is thoughts (*noemata*). And these are one by succession like number but not like [continuous] magnitude." For how could a magnitude think, with what parts or points (*On the Soul* 407 a 8)?

So while sensing is of magnitude, thinking is related to number. But recall, time is a number counted up by the soul. Here Aristotle's account passes suddenly from the firm ground of reductive reason to the deep waters of a perennial enigma: How does the soul count motion, which is a continuous magnitude?

2. *Motion*

For present purposes, Aristotle says, it does not matter whether we speak of motion or change (*kinesis, metabole*, 218 b 19). What then is motion in the largest sense, as Aristotle understands it? Motion depends on two aspects of any being, its capability or "power to come to be" (*dynamis*) and its fulfillment or "abiding at its own end" (*entelecheia*). Natural beings have the power to come to be what they were meant to be. When they are so becoming, whether by moving to their proper place or by changing through growth or other alteration (which usually involves some change of place), they are in a peculiar semistate, a state of transition, in which their power to come to be has come into its own and is fully active. Motion is the fulfillment of a being that has the power to come to be, insofar as it has that power (201 a 11). A mobile in motion is therefore at work on becoming fully itself. It is, one might say, fully at work in becoming fulfilled.

This bottomlessly profound view of motion and mobility is, however, largely set aside in the inquiry into time. For in that context what matters is only a colorless property motion always has—though it has it not merely as having so much duration but more intimately, in itself, because it is *sensible*. Hence time measures both the lasting of the motion and the motion itself (221 a 6). Thomas Aquinas explains Aristotle's assertion: *Mobiles* are measured by time (which is a discrete measure of magnitude) in respect to the duration of their state but not with respect to their [qualitative] being, while *motions* are measured by time in their very being, as we would say, in their extension (601). *Extension* is the colorless characteristic through which motion becomes temporally measurable.

The emphasis on the measurable dimension of motion coincides with the emphasis on motion in place, where "before and after"—and hence the paths they direct—are said to be found first of all (219 a 16, 223 a 28).

3. *Before-and-After*

It is essential for preventing the analysis of time from being circular that "before and after," to which the time-number is referred, should not be primarily temporal terms, and that is why Aristotle points out their spatial primacy. He had of course a much more powerful derivation available for these terms, but its suppression is consonant with the de-emphasis in Book Four of the mobile and its qualitative properties just mentioned. For the progressive becoming of a natural being has stages that appear in time but are really prior and posterior in thought. Now the counting that will give the time-number needs an anchor and a direction. So one would expect Aristotle to choose the direction-marker—for that is how before-and-after functions—that is most deeply embedded in the mobile. And indeed in the *Metaphysics* such a priority is formulated (1028 a 24).

In the *Physics*, however, where locomotion prevails, it makes sense that "before and after" should not be *primarily* moments in the mobile but locations of place. For place too has a capability and power that it imparts to bodies that then go to the place proper to them (208 b 11, 211 a 6). Hence it has determinate before-and-after, and so motion can be measured and numbered without loss of orientation. I leave aside here the puzzle how distance measurements can be performed in "place," which, as Aristotle understands it, is emphatically not a dimensional notion (*mete diastema*, 212 a 4, p. 48).

In either case, "before and after" might be construed as prior and posterior. (The words are the same: *proteron-hysteron.*) But in this book of the *Physics* they are primarily position-terms; in motion itself they are derived by analogy from magnitude, magnitude understood, however, primarily as a path-length. And so they descend to time. In fact, Aristotle goes so far as to say that before-and-after *is* the motion, though separable in thought (219 a 20 ff.). That must mean: motion has, or even is, directionality.

I should say here that in the discussion of the soul there will be good reason to translate the pair as time-terms after all, as "earlier and later" (p. 47).

Aristotle reinforces this second prime property of motion (the first was magnitude) by an observation: "Before and after" are spoken of oppositely round about the Now. In the past, "before" is further from now than "after"; in the future "before" is closer. This asymmetry around the present betokens the directionality of motion and time, for it means that the series goes: before → after → now → before → after, progressing steadily forward—just as does numbering.

4. *Number*

Aristotle wants the reader of the time-chapters to understand what number is. The central chapter of the inquiry (xii) begins with its relation to the unit (220 a 27 ff.), while a significantly placed coda deals with its own nature (224 a 3 ff., related to 223 b 13 ff.). And certainly, if time is a number (221 a 14), the meaning of number for Aristotle becomes central, and the more so if numbering is a very deep function of the soul, as I shall claim.

Number is the higher branch of quantity (*poson*), the discrete branch, called multitude (*plethos*). When such a multitude of indivisible ones is finite and thus numerable, it is a number (*arithmos*, *Metaphysics* 1020 a 13). Since one is not a substance, the numbers composed of ones do not have a separate nature of their own (1001 a 26); they do not have the substantiality of bodies, nor are they prior to them except in reasoning speech (1077 b 1) that derives them by "abstraction" (*aphairesis*, literally, "lifting-off," e.g., 1061 a 29, *On the Soul* 431 b 13 ff.). As for all quantities, the most defining characteristic of numbers is to be "equal or unequal" (*Categories* 6 a 27).

Aristotle especially wants us to keep in mind that:

(a) The one or unit (*hen, monas*) is not a number (*Physics* 220 a 27 ff.). It is the principle and beginning of number and, so to speak, sets its tone—the unit tells what the number is a number *of*. When the unit measures lines or other continuous magnitudes, there is not a smallest unit-measure, but there is a smallest number of unit-measures, namely two. When Aristotle analogizes the Now and the monad (220 a 4), he is making a leap, not quite legitimate, from length-measure to point-principle.

(b) A number is, at least to begin with, a multitude of *things*: a heap of stones or a pack of dogs or a flock of sheep (224 a 3 ff.). Such concrete numbers are each counted by their own unit, a stone, a dog, a sheep (223 b 14). These different concrete number-heaps, numbers in the primary material sense, may have the same or different numbers in the disembodied and abstracted mathematical sense because multitudes are by their nature equal or unequal; so when the mathematician's "number simply" (*arithmos haplos*, 220 a 27) is counted the former pack-number may be indistinguishably equal or legitimately unequal to the former flock-number.

(c) Finally, motion and time can measure and mark each other reciprocally (220 b 15 ff., 223 b 16). For if we count up units of (equable) motion over a path we get the time-number, and if, reciprocally, we have the time-number the length of the path is determined. Aristotle illustrates this reciprocity from the double nature of number as embodied and abstracted. We know the multitude of horses by an abstracted number (which operation corresponds to the time-number gotten from the motion-path). And reciprocally, if we know the "abstracted" number of the troop we can recover the concrete multitude of horse-units (which operation corresponds to getting the concrete path-number from the time). The reason that these arguments for the reciprocity of time and motion are important is that they will serve to justify clocks (p. 47).

Note that Aristotle has much to say about the Being of number, its ontology. But he has little to say about the activity of counting, its generation. And yet that is the ultimate question concerning time in his understanding, for there is no time without counting (223 a 23). Number has two meanings; it is the number counted (*arithmoumenon* or *arithmeton*) given a countable multitude (objective number, we would say), and it is the number "by which we number" or count (subjective number). Of these, time is emphatically the former (219 b 8). But the more originary number, the one that arises from counting—of that we hear very little in the inquiry on time. And yet one might say that such counting, this human activity, more than establishes the number of motion—it encompasses it. For it goes on when there is no motion to count (221 b 9). This, the soul's counting, will be taken up below (p. 46).

5. *Now*

"Now," the crucial term in Aristotle's inquiry into time, gains its significance from prior notions. Into it go: continuous quantity or magnitude, motion directed by before-and-after, discrete quantity or number. For the Nows function to cut the magnitude of motion successively and to make the numbering of motion possible.

As magnitude-cutter, particularly of locomotion, the Now is conceived on the analogy of the point of a line, for it acts approximately pointlike: It is indivisible (233 b 33). It connects the past time (or motion) with the future and is thus the very "continuity of time" (*synecheia chronou,* 222 a 10). On the other hand, it cuts time up, marking off time (or motion) into lengths to be counted. Thus it is a divider "potentially" and unifier actually (222 a 19). The Now is always the same in thought, as points are indiscernibly identical in themselves (219 b 19), and it is always different in temporal position, as points have different positions locally. Thus the Now is doubly double: It both divides and unifies, and it is both same and different. Like points, Nows don't add up to make structures of time (*On Indivisible Lines* 971 a). Furthermore, if one mixes conceptions of physics and mathematics, just as a point may be said to generate the line as its head, so the Now is, as we would say, at the leading edge of time. This model can even be said to picture the factuality of the past as against the indeterminacy or nonbeing of the future: the line already drawn and the line not yet realized (p. 172).

That is the Now driven to mathematical abstraction. But the Now has a much more pivotal function: It is the tangent point of marking thought and passing movement and the intersection of receptive sense and moving thing.

In yet other words, the Now is the movement-moment of the *mobile* (*to kinoumenon* or *to pheromenon*), when the moving thing is *present* to sense, for "sensation [is] of the present" (*tou parontos, On Memory and Recollection* 449 b 14). And with sensation comes knowledge, for the Now is closest to apprehension and "especially well-known" (*gnorimon malista, Physics* 219 b 29). It is when the soul says "now" twice, one Now being before and the other after, that we speak of *a* time (a 29). Thus "it is by the mobile that we know before and after in motion" (b 25), for the mobile is not only the source of change but also the captor of our attention to change, and so to time.

In short, it is the mobile that offers the soul a presence on which its now-saying can successively alight. Aristotle summarizes these relations, touched on above, by this extended analogy: As motion is to time so is the mobile to the Now, and so also is the number to the unit (b 24, 220 a 4). The first relation is that of extended passage to its measure, the second that of the presence of a thing to sense, the last that of the abstracted result of counting to its element. It is that last that will be most perplexing.

The mobile, it must be said, is the only substance, the only definite something (*tode ti*) in the many-linked story of time. And yet it too, like the Now, may in this context be regarded as similar to a point (219 b 18).

Now when Ptolemy asks for the earth to be regarded as a point, he gives as a justification its fantastically small ratio to the heavens (*Almagest,* I 6). And when Newton treats bodies as points in dynamic systems he is relying on the provable notion of a center of gravity. But what is Aristotle's excuse for treating a body as a point? His motive is plain enough: to preserve the continuity of time, which requires a point-Now, in the face of the fact that such a point-Now can alight only on a similarly nondimensional mobile. But his justification, if it is meant as one, is weak. As has been said (p. 41), motion is measured by time in its very being (221 a 6), while a mobile is not essentially quantitative. So while time measures, as Thomas puts it, the changeable "existence" of the mobile, that is, its being in time, what it is in its essence is not in or of time but corresponds to an atemporal Now (602 ff.). In fact, Thomas plays a little riff of his own on this idea: If all changeable properties are removed from the mobile, what remains is the self-same substance; it now comes to a stand, and this stable being enters eternity. But this standing Now of essences is far afield from Aristotle's theory of motion, even from the everlasting motion and atemporal movers set out in the last books of the *Physics*.

In any case, an argument to the effect that the mobile meets the Now insofar as it is a magnitude and not a substance would not elucidate how sensing can take place if the mobile presents itself as a point marked by a point. By this account sensing becomes very dubious. Or rather this account emphasizes a perplexity that recurs in other Aristotelian passages about sensing.

In *On the Soul* it is said of the common sense, the sense that receives and discriminates time and motion, that when it "takes in and senses" (*noei kai aisthanetai*) something, it says *now* that something is *now* the case, for it is itself undivided and judges in an indivisible time, that is, in the Now (426 b 22 ff.). So the common sense would seem to be punctual, so to speak.

On the other hand, in *On Sense and Sensibles* the opposite argument is made: We sense the continuous magnitude of all sensibles continuously in continuous time and "not by any of the Nows in it" (448 b 6, emended).

This is not a brute case of self-contradiction but the expansion of a perennial perplexity: How long is the present? By "present" I here mean that moment, which Aristotle calls "now" in the *Physics*, when the world is before the soul and they touch each other. That Now of presence is too short for sensation if it has to accommodate itself to a world that is everywhere magnitude, or, as we would say, is extended. The perceptual present, on the other hand, which we will meet later as the "specious present," the dilation of the Now, is repugnant to the mathematical model of time that reason extracts from experience. I have heard of no really satisfactory resolution to this problem (though Husserl will attempt one, p. 141; Gallagher, 17 ff.). It therefore deserves to be classed as a mystery: the mystery of presence in time (p. 203). It is clearly related to the perplexity concerning the soul's counting of magnitude. These are the problems Aristotle sets up for our contemplation, and they will haunt all later time-inquiries.

6. Soul

Soul (*psyche*) is Aristotle's word—at least in the inquiry concerning time—for all these human powers together: taking-in or thought in the widest meaning (*noein*), sensing (*aisthanesthai*), knowing well or recognizing (*gnorizein*), marking or bounding (*horizein*), and finally judging or discerning (*krinein*), which is thought and sense together (*On the Soul* 432 a 15). These are the powers by which we come together with the world and discriminate its presentations.

It is impossible for time to be if there is no soul, for soul alone can by nature count, and of the soul it is the part that engages in "thought" (*nous*, *Physics* 223 a 26 ff.) or "thinking" (*dianoia*, 218 b 22) that does the counting. The lineage from soul to time seems to be this: Our soul senses Nows (219 a 31), for without the Nows there is no time (219 b 33), just as there is no time without the Now. And the Now, alighting on the mobile, starts (or continues) the count by functioning, along with its sensing of the mobile, like a unit (220 a 4), a unit-beat or mark, so to speak. The accumulation of units, the resultant number, belongs to motion and *is* time.

But if the count belongs to motion, the counting belongs to the soul. And since an uncounted number is an impossibility, for number is something counted or countable (223 a 25), *the soul must be said to originate time*. Though Aristotle never quite comes out with it, it must be so.

Now here is a strange additional fact. The soul counts not only motion but also rest. Even when there is no sensing, when the body is unaffected, as in the dark, there *is in* (*enei*) the soul "a certain motion" (*kinesis tis,* 219 a 6). To be sure, in *On the Soul*, rest is said to be a common sensible that is sensed by the absence of motion, as number is sensed by the negation of continuity (425 a 19). On the other hand, in the time-section of the *Physics* Aristotle says that when nothing happens in us between two Nows, as in a loss of consciousness, they merge and there is no time (*Physics* 218 b 18 ff.; rest without respect to its apprehension is discussed in 238 b 23 ff.).

So the conclusion has to be that time elapsing (1) does not depend exclusively on external motion, on the motion of mobiles, but (2) does depend on some sort of internal or psychic motion.

Is Aristotle equivocating when he says things like: "But there is no [time] without change, for when we ourselves don't change our thinking, or our changing escapes us, there does not seem to us to be time" (218 b 21 ff.)? Yes, for it was physical motion and the mobile that were at first the seat of the before-and-after, but now that motion is of or in the soul.

This second motion, this mystery motion, is actually touched on once more in *On the Soul*: Neither "paying attention" (*phronoun*) nor sensing should be thought or spoken of as psychic change as change is used of nature, for—speaking summarily—these activities are more like fulfillments, while change and motion are only on the way. So what goes on in the soul is "another genus of alteration" or "another species of motion" (417 b 8, 431 a 7).

There are some clues about the character of this psychic motion. The soul is emphatically not a magnitude. Thus it is not a body "to be moved by nature," that is, as the mobiles of the *Physics* are moved. It is not even itself nature, in the primary sense, because nature is that which has "within itself" a principle or source of motion (*Physics* 192 b), while the soul does not have but *is* such a source of motion in bodies and therefore the cause of *their* being natures; indeed it is the primary source of motion in place. But the source of motion does not itself have to be moved. It is the appetite by which the soul is the source of motion for its body, not by an extended part of itself (*On the Soul* 403 b 26 ff., 405 b 33 ff., 415 b 23 ff., 433 a 31 ff.). This fact—that the soul is not a continuous magnitude—gave rise to the perplexities of the previous part (p. 45) about sensing and its continuity or discontinuity. These difficulties are really a primary version of the perennial problem about mind marking matter.

But if Aristotle wavered a little about the continuity of the activity of the particular senses, there is no equivocation about thinking and discriminating. One telling passage has already been quoted: The discriminating sense "says now and that [something is] now . . . so that the discriminating is partless (*achoriston*) and in a partless time" (426 b 29).

The mind is not a magnitude, although—and this is a crucial passage—"thought (*nous*) is one and continuous, just as is thought-activity (*noesis*), and thought-activity *is* the thought-objects (*noemata*)." Aristotle does not here mean the continuity of extension but of internal unity. "These, however, are one by succession, as is number but not as is magnitude. . . . For how could [thought] think, if it were a magnitude?" (407 a 7 ff.). And again: "The motion of thinking is not something continuous and underlying, as is that of mobiles" (*On Indivisible Lines* 969 a 1).

In other words, thought is something very strange, something like a continuous number, while thinking is like counting. Counting, finally, is defined—quite incidentally—as "thinking touching [members of a series], each one singly." And it occurs "with pauses" (*meta epistaseos,* 968 b 3, 969 b 3). Of thought-activity it had previously been said as well that it is "more like a rest or a pause than like motion" (*On the Soul* 407 a 33).

Thus, the soul, in its highest reaches of fulfilled thought (*nous*), is akin to number. Of "human thought" (*anthropinos nous, Metaphysics* 1075 a 7 ff.), Aristotle says that it, as does anything put together with matter, "stays in time." In *On the Soul* he shows just *how* thinking has temporal parts (430 b 7 ff.). Time is counted in a soul that is, by reason of its body, in time.

Hence in that thinking, or better in that "thinking-through" (*dia-noia*), which is closer to sense, the soul can alight on the elements of a physical succession, such as motion, touching each now-member. But when necessary it can count without a body or path or number-heap being present. What could it be counting and noting but the simple Now, and "that it *is* now"? The soul is, in other words, counting its own marker of presence. This activity comes close to the pulsing of pure—and here I mean empty—self-awareness: originary time (p. 205). But that means that "before and after" need not always be physically given. *Proteron-hysteron* may sometimes mean "earlier and later"—by the soul's own time.

7. Clocks

Let me descend from the most speculative to the most workaday aspect of time: its measurement. For Time embraces time, and we may think of Time and also of a span of time. Both have the same name, *chronos*.

Now Aristotle does not speak of sundials or waterclocks, but he tells us what makes them possible: It is the fact that as time measures motion so motion can measure time, as set out earlier in this section (p. 43).

Aristotle's prescription for determining comparative speeds of motion—it might pass for a definition of speed—requires the mobiles to be traveling equably, not necessarily in a straight line but over the "same distance" (223 a 2); the one that arrives before the other is faster. Aristotle says nothing about clocking the arrival times, that is, counting the motion-number so as to get a quantitative ratio of speeds. Moreover, as far as I can tell, time is not taken as the independent variable in those books of the *Physics* that were particularly given the title "On Motion" in antiquity, namely V–VIII. Nor is a varying rate of change, that is, acceleration, considered, since Aristotle lays it down that there can be no change of change (225 b 17). Perhaps both of these, time as independent variable and acceleration, become thinkable only when there are portable time-pieces.

Yet another observation: It is often said that Aristotle has no conception of the infinite, homogeneous, merely dimensional, Cartesian coordinate space with which we so familiarly deal. What he expounds in the *Physics* is *place*, the limit of a "containing body" (212 a 7), conceived of as itself seated in a cosmic plenum. It is "where" a thing is not as located by distances but as contained in its cosmic home. Thus Aristotle deprecates the understanding of space in terms of distance (*diastema*, 209 b 7) as falsely treating place as formless matter. But that denial does not prevent him from using the term "distance" for spatial removal, as above he used "interval" (*apostasis*, 223 a 8) for time elapsed between Nows. Somewhere there is lurking here a notion of mere dimension, or else how could motion-paths be measured?

Yet one more reference, possibly unnecessary, to a latter-day notion missing in Aristotle's theory of the measurement of time: real numbers. All the quantities measured—paths (distances), motions (the activity that covers distance), and time-spans (the motion between nows)—are continuous, but their measure, time, is a number and discrete. Aristotle's theory does not suffer in the least from the absence of a number-continuum, that is, real numbers—as does no later theory of measurement. For all human measuring takes place in a world of commensurable quantities, meaning quantities that have a common countable measure. The unit-measures may, after all, always be taken small enough to exhaust the dimension measured, because of the limits to our discerning power. So Aristotle's real problem still is just the one pointed out before: How does the counting soul come together with the magnitudes of the world?

There is a second serious problem, however, in respect to one more notion that is necessary to the measurement of time: simultaneity. Unless one can persuade oneself that it is one and the same time everywhere, clocks will have a strictly local use; they might even work for just one motion's time. (I am of course talking about the Aristotelian case of a well-contained cosmos, to which relativity concerns about light signals are not pertinent.)

Though Aristotle's treatment of the compatibility of different sorts of change *seems* to conclude that to be comparable with respect to speed, motions must be of the same kind (249 a 2 ff.), the phrase "in the same time" is nonetheless used of diverse changes, as is only sensible. That is one step toward making simultaneity possible. For besides the requirement that motion be able to measure time, clocking requires, as was said, that all the motions be referable to some third, standard, motion. This may be what a chapter on different sorts of motion in Book Seven (iv) tacitly assures.

The trouble is that in the inquiry into time of Book Four, (1) each mobile or motion is timed on its own, and (2) the argument for one universal time, that is, for simultaneity, is only feebly made, though it is forcefully enough asserted—as the large number of references shows (218 b 13, 219 b 11, 220 b 6, 223 a 24).

The last of these references suggests a solution to the quantitative comparison of motions: The units may be qualitatively different, as they are, for instance, in locomotion and alteration, but their count is the same, in other words, the monads may be different, but the number the same (p. 42).

Does this explanation by itself solve the problem of how "all time is the same simultaneously" (*hama*)? Only if either one soul is counting all motions everywhere or if all psychic counting is synchronized—neither of which is plausible.

Yet something like the soul being everywhere seems to be implied in Aristotle's insisting that it is one and the same Now that marks all motions (223 b 3). When he had spoken earlier on of the Now being one and the same, he had meant that it always has the same character, that of dividing earlier from later (219 b 10 ff.). But here he says explicitly that there are not two different times running simultaneously. And yet the Now is *one* soul's marker.

Shortly after and near the end of the book he finally comes to what may be a solution: a standard time indicator, a clock. We are so used to portable analog clocks with round dials that we might forget that besides the sundial, which was more than half the time out of commission and not perfectly equable, the ancients had but one such clock—not portable, to be sure, but universally present: the rotating heavens. The heavens differ from a clock mainly in that the marked dial itself moves rather than the hands. Now if all souls are willing to entrust their counting to this motion because they trust it to be equable, and if they read the position of the sun by day and of the stars by night, then all human beings can have the same Now (setting aside soluble technicalities such as the sun's seasonally varying ascensions and different time zones, which do not concern Aristotle). Measuring time may be a highly localized activity, and yet clocks may tell time for everywhere as well as for right here. But without reference to a heavenly clock, simultaneity seems unachievable in Aristotle's theory.

And he does finally say that the rotational motion of the heavens is rightly preferred for measuring time, because it is equable (*homaleis*, 223 b 22) and because of the periodicity of human (and other) motion: "Nature wants to measure births and deaths by the numbers [given by heavenly bodies]," though it is prevented from doing it accurately by the indeterminacy of matter (*Generation of Animals* 778 a 5 ff.). Moreover, the heavens' motion in place is primary and perpetual (*Physics* 261 b 27 ff.). Aristotle has already taken care of the possibility that a heavenly circular clock might be construed as showing cyclical rather than progressive time. For while each motion has its proper, "primary" time (236 b 21), that time is "embraced" by a greater time (221 b 32); it never ceases but is always at the beginning (222 b 8). And this ongoing time never repeats itself: two days are the same time span but not the same time, since the change that is past differs from that to come (220 b 7). Now these arguments only hold if changes are never indiscernibly identical, or (and that is probably what Aristotle *would* say) because of the aforementioned indeterminacy of matter that ensures perpetual contingency. So time is progressive, but its parts can be clocked periodically.

Is it utterly clear that Aristotle ends up so far from those Pythagoreans who identify the heavens and time, the ones he began by criticizing as naive (218 b 1 ff.)? Their problem of explaining the unitary and embracing character of time arises for Aristotle as well and is resolved by them and by him similarly through recourse to the heavenly revolutions.

I wonder if Aristotle would have been willing to see the various Pythagorean theories as prefiguring in an imaginative way elements of his own thinking: In making the soul encompass the heavens, they acknowledged that time is generated by soul (p. 6), and in making the heavens act like a clock they solved—as he thinks, too naively—the question: How is the world in time and in *one* time?

III. Time

When Aristotle closes his inquiry (*skepsis*, 224 a 17) into time, the willing reader may come away with two very different impressions. One is that Aristotle has in a revolutionary way evaporated time—has shown it in analysis to be completely dependent on elements that are themselves insubstantial, whilst reducing it to the sparest of definitions: a number attaching to motion insofar as it has succession built into it. The world of nature consists of "physical" magnitudes in directed measurable motion, and the measures accumulate; such accumulations are called numbers.—There is not much to that; time has been reduced to something soberly secondary.

The other impression is that in saying that time is the number of motion and in coming to grips especially with the "Now of the present" (222 b 8) as the moment of counting, Aristotle discovers in the soul's numbering a different and deeper origin of time, but one beset by long-range perplexities.

Heidegger says in *Being and Time* that the source of time does not become a problem for Aristotle (81). He trivializes Aristotle's understanding of time by a misstatement and a mistranslation. He takes the Aristotelian definition to mean that "time is what is 'counted.'" And that is of course false: primarily it is motion that is counted. And he translates "number" in Aristotle's definition as "that which is counted." That interpretation of course misses the duality of number that Aristotle is at pains to point out and that leads him to the soul's activity: Number, *arithmos*, is *both* what is counted *and* what is *countable;* it is not only a result but the expression of a capability: the capability that moving substances have of being marked and noted by the soul, by us. Moreover it is not really, as Heidegger says, the Now that is counted, but the measure or span marked by the Now. Aristotle's time is not, as Heidegger claims, "now-time," but had better be called "noting-time." The Now functions not as a mere pointer but as the captor of a

Being. Nor can it be accurately said that Aristotle has a merely "natural" understanding of Being; the soul is, strictly speaking, *not* natural (p. 46). In one respect though, it seems to me Heidegger is eminently right: The Now marks a present and a presence; the soul says "now" and also that "something is now." Present presence, the moment when physical substances and sensing intellect intersect—these are central for Aristotle the inquirer. But then there is also Aristotle the contemplator, for whom temporal present and physical presence yield to the atemporal identity of thought-activity and thought-object. In any case, as I said in the beginning of this chapter, Aristotle's apparent reduction of time is fraught with perplexities that are philosophical openings. I do not just mean that his account gives rise to specific perplexities, but something much more serious: that a reductive understanding of time leads straight to the deepest questions about the connection of human noting or marking to the sensed world.

It is a fair question why Aristotle so strongly emphasizes the reductive resolution of time-perplexities over the problematic remainder. One answer is that that is his way, and that he says what he can and leaves what he must. Perhaps a better version of this answer is that he tells the main truth: Time in the world is scarcely anything; the rest comes after. But the best answer is, I think, that this downgrading of time fits neatly into his view of divine, natural, and human affairs: The *Physics* ends with a proof that the First Mover of the world is not in motion and is not a magnitude (267 b 18), and in harmony with their First Mover, natural beings themselves are moved not by the force of time but by the end in which they are fulfilled. As far as human beings in particular are concerned, the book where we would expect to hear about human time, the *Nicomachean Ethics*, ends with the praise of contemplation as the highest and the most peculiarly human activity, an activity that is, if in time, only incidentally so. And before that climax nothing much is said of the temporal aspect of human life. For a human life (*bios*) is conceived as an organic development. It is *not* a time span, or a diminishing ratio of spent to remaining time, or a location in human history, or the inescapable labor of breasting a future and digesting a past and exploiting the present.

IV. Phases

Past, present, future are the "phases" of time. "Phase" is not a word Aristotle uses. Nor can Aristotle use nouns for the three phases, since Greek has none. The most characteristic of his various usages is the neutral plural (or singular) of numerous process verbs: "[the things] that have come about, have passed away, have gone by," and "[the things] that will be, will come about, are about to be." In all of these usages for past and future the words speak of tensed *things* or happenings, not of time-phases. The Now, which is not time but only incidental to it (*symbebeken*, 220 a 22), is not really a phase; I don't believe the present is ever named in the *Physics* except adjectivally as, for example, in the "present change" (*metabole parousa*, 220 b 7). In short, both Greek itself and Aristotle's use of it discourage a substantive view of the phases of time. And in the inquiry on time they play no role—a remarkable fact.

To read about the past, not past events such as are spoken of in history, but the psychic grounds on which the past comes to be, we have to go to the wonderful little treatise *On Memory and Recollection*. Here it is established—an observation that will turn up from Augustine to Wittgenstein—that the present (*to paron*) belongs to sensation and thought, the future (*to mellon*) to expectation and conjecture, and the past (*to genomenon*) to memory (*mneme*). So, strictly speaking, we know (*gnorizomen*) only the present; moreover these cognitive modes are never to be mixed, so that, for example, we cannot remember the present while it is present (449 b 14), that is, the Now *in* the Now (b 26). "Hence all memory comes with time," and only animals that have sensation of time can have memory. It follows that memory is not actual sensation or knowledge (*episteme*) "but a sensation or affection when time has passed," in which "one has knowledge and sensation but without their being at work" or actualized (*energeion,* b 19).

This strange condition of inactual sensation and knowledge is met by the imagination; whatever is remembered is in its nature also imagined (450 b 24). The imagination-memory is like a painting, produced as by a seal impression (a 32). It is through the analysis of what it means to look at a picture that Aristotle explains what happens when we remember: We experience the internal picture in a double way; the experience is at once a present *affection*, as of a quasi painting, and a *memory,* as of a reminding likeness; we now see the inactual image-person and are reminded of the actual person in the past mode. This analysis prefigures Husserl's interests set out in his *Phenomenology of Internal Time-Consciousness* (p. 145).

The point of this inadequate sketch of one of the most interesting psychological passages written in antiquity is to show that for Aristotle the past is inactual presence. One consequence is that history, which deals with inactual particulars, is less serious and philosophical than poetry, which deals with universals unaffected by time (*Poetics* 1451 b 6).

We might expect memory to be specifically related to the inquiry into time. For time arises through psychic counting; how then does the resultant time-count come about? Is it memory that synthesizes the succession of monads into a counted whole? Aristotle does not say; it is an outstanding perplexity.

The problem Aristotle does treat, and in fact regards as the "greatest" issue concerning recollection (*anamnesis*), that is, the deliberate recovery of memories, is how lengths of past time, measured or indeterminate, come to be recollected (452 b 7 ff.). He postulates that we have something in us—probably the common sense—that allows us to discern greater and lesser times. Moreover, he thinks it reasonable that time and distance should be measured similarly. Distance-magnitudes are compared through "proportional motions" of the soul. The soul contains figures and motions similar to the external figures, though "internal" (*entos*) magnitudes are usually considerably smaller than their sensory counterparts.

Now this explanation of internal distance-measurement surely does *not* work for time, as Augustine will point out (p. 119). For while it is possible to inscribe into imaginative memory-space similar figures, the time-sense, as Kant will insist, is unidimensional, and it is impossible to view time-stretches side by side, in fact impossible to view them at all (except by representing them spatially, which begs the question). And since Aristotle rightly observes that memory is at work (*energei*) when the psychic motion related both to the thing (*tou pragmatos*) and to the time have come together, that is, when we become aware of a thing and also *that* and *where* it was in past time, he cannot give an account of a temporally coherent internal past. That too is an outstanding perplexity; Husserl will address it exhaustively.

If the past is inactual, so much the more is the future. The future is considered in *On Interpretation* with relation to propositions (18 a 28 ff.): The past, inactual though it be, is certainly determinate, and particular propositions about it are either true or false. Not so for the future. Particular propositions about it are *now* neither true nor false, for both chance and human deliberation introduce the possibility of any proposed fact coming about or failing to come about in the future (19 a 7 ff.). Logicians have found in these passages tangles for analysis (Hintikka, ch. VIII). Yet the main point is plain: The future is not fixed by necessity (p. 174).

Recall, moreover, that Aristotle does not disagree with the folks—of a conservative cast of mind, one imagines—who think of motion, and so of time, as primarily destructive: *ekstatikon*, "displacing, *deranging*" (*Physics* 222 b 17), is the word that will be put to so different a use by Plotinus and Heidegger (pp. 98, 103). But recall too that this displacement is not actually a force of time, but a characteristic of motion. Hence the future is inactual, indeterminate, and derivatively deranging. I am not claiming that Aristotle, who gives such serious accounts of the practical life and of the role that deliberation, choice, and prudence play in life, has no interest in the futural aspect of human time, only that the future is not for him a focus either of fear or of fulfillment. It holds no being that does not develop from the past or the present, modified by chance and choice.

B. KANT

Once again the exposition of a theory of time, now Kant's mind-boggling inversion of the tradition, is intended partly for its own sake and partly to reveal its limited but deep affinity with Aristotle's understanding. This affinity is deep because it goes to an ultimate question to be asked of time: How can human beings be thought to sense time? The deceptively simple shared answer is: by counting.

That this affinity is, in fact, also limited scarcely needs reinforcing. Kant, after all, inverts Aristotle's world of a multiplicity of individually formed substances that each have Being into one universal substance that is an appearance: *substantia phaenomenon*—matter (*Critique of Pure Reason,* 333; simple numbers refer to the second or B edition of 1787; the prefix A refers to the first edition of 1781). Kant's most specific criticism of Aristotle, expressed together with the respect for his still usable work, does not, however, concern this-worldly substances but Aristotle's extension of principles applicable to sense to the supersensible realm, that is, to a senselike intellect ("Of a Newly Arisen Refined Tone in Philosophy," 1796). It is to the eradication of such an unwarranted widening of knowledge beyond that gotten by discursive, categorical thinking on the basis of sense that the *Critique* is devoted (26, 783). A myriad of differences in detail between the two authors will be taken up in the course of the comparative exposition below.

And yet there are plenty of continuities and comparabilities as well. Kant calls the *Critique of Pure Reason*, on which this chapter will be largely based, a "propaedeutic," that is, a preeducation, toward an organon (25). Aristotle's logical works were called "the Organon" in late antiquity, meaning "the instrument of inquiry." Kant's organon not only explicitly borrows much from Aristotle, for example the concept of a set of categories or predicates of judgment (105), but is, in a world-inverting way, just as much an ontology (873), an account of Being—though one that does without individual beings. Both works have a preparatory purpose, though in antithetical ways: Aristotle also thinks that the inquiry into Being is in want of a propaedeutic, but he means by that a proper personal training in logic (*Metaphysics* 995 a 12), not a critical inquiry into the faculties of knowledge. The examination of the objects of knowledge comes *before* the inquiry into the soul (*On the Soul* 415 a 21).

In sum, Kant introduces into the philosophical tradition a true novelty, a radical and systematic inquiry into the sources of knowledge: epistemology, an account not of the objects but of the faculties of cognition, a study into which Aristotle (one may imagine) would not have wished to follow him. What seems to me most revealing of the difference between the two is the question that each of them regards as unaskable, as absurd. For Aristotle it is the question whether there *is* nature (*Physics* 193 a 4); for Kant it is the question how a thinking subject can have intuition (A 393). This unaskableness marks the ultimate given for each: the being of nature for the one and the constitution of the soul for the other.

Kant establishes in the *Critique* a new version of a human being. For Aristotle a human being was a *nature* among natures, and the human soul, though not itself a nature, was the working principle of a natural body. For Kant, a human being is primarily a system of *spontaneity* (meaning the ability to produce representations, though empty ones, on its own) and of *receptivity* (meaning the capability of being affected externally), cooperating to give rise within itself to a system of law-governed nature.

Nonetheless, the internal constitution of the Kantian soul is borrowed from Aristotle's triune scheme in *On the Soul*: a two-tiered intellectual faculty, Reason and Understanding (analogous to *nous* and *dianoia*), the middle function of imagination, and the low but indispensable sensibility.

In the following exposition, I shall be compelled now and then to shuttle back and forth between similarities and differences so that certain essential features in the two theories of time may the better stand out as identical or analogous. Let me give a preliminary sketch of topics on which the theories coincide, or at least in which the issues that preoccupy the two philosophers are analogous. (1) For both, time is related to, and perhaps *is*, inner counting. (2) For both, sensing provides the pivot point of time. (3) For both, the imagination, its obscurity notwithstanding, is crucially responsible for the phases of time. (4) For both, the relation of time to space and to motion is central but unresolved.

To extract a theory of time from the *Critique of Pure Reason* means to make a choice among its many aspects. The word "critique" has at least three meanings in Kant's work. One, already mentioned, is the criticism of the human attempt to take knowledge beyond its warrant, beyond the proper "limits of reason." In this sense, critique, in reining in the excess of pure reason, is antidogmatism (424, 766, *Prolegomena to Any Future Metaphysics That Might Claim to Be a Science*, Scholia); dogmatism is metaphysics ungrounded in critique. The term is next taken in a positive sense: the analysis of the cognitive faculties themselves, together with the demonstration that they must be just as Kant sets them out to be, if experience (meaning a significant knowledge of nature) is to be compelling, that is, necessary and universal; the result is a new grounding of metaphysics (873). The critical enterprise in the third sense poses its project in terms of a technical question: How are a priori synthetic judgments possible (xvii, 19)? This question demands a warrant in our cognitive constitution for judgments we make "in advance of" (a priori) any sensation that might come to us, and that nonetheless contain predicates that add something new to the concept containing the matter of the judgment. In other words: How can we, from the very beginning, find in ourselves significant knowledge that is neither sensible nor supersensible?

In the course of analyzing and delimiting human knowledge Kant demarcates a region in which the human being does indeed have access to the supersensible, a realm of spontaneity in which action can occur that is not bound by the laws of nature but is originary and free; this is the territory of morality (832).

Accordingly, three terms are used of the human being. (1) It is a *person*, a free moral agent acting in accordance with self-given law only. (2) It is a *subject*, an unknowable "I," underlying all its cognitive functions but most immediately its thinking (278, A 350, 404). (3) Last but not least, it is a soul (*Gemüt, Seele*), a being that has a sense of itself and appears to itself in an "inner sense" (33, 37, 400, 874). It is with respect to the inner sense that critical philosophy shades into psychology (*Anthropology,* 24). But this inner sense *is* time.

It was to locate time in the critical enterprise that I engaged in this long preliminary. Time will be most approachable if the *Critique* is regarded as a new founding of human nature whose center is time. Along the way I will be recalling that something similar might be said for Aristotle: Time is nothing in itself and yet points to a human crux.

This is the place to mention two interpretations of Kant's theory of time that are antithetical in outcome but identical in their starting point, which is the shared realization that time is the deepest, or perhaps better, the most intimate feature of humanity for Kant. Hermann Cohen, in his *Logic of Pure Knowledge* (1902), feels compelled by the deep-rooted nature of time to correct Kant: Time is not an inner sense but a category of the understanding, a classification for which Kant criticized Aristotle (107). Heidegger, Cohen's latter-day successor at Marburg, is driven by the depth at which Kant locates time and by his own agenda to see in it the root of *all* faculties, the ultimate root of human Being, beyond mere thought. In *Kant and the Problem of Metaphysics* (1929) he claims that Kant came close to seeing where his radical critique was leading but turned away as before an abyss (para. 35).

Details in Kant's texts speak against this further radicalization of his already radical insight (p. 88 f.). I mention Heidegger's treatment here not only because it is the most original interpretation of the time-passages in the first *Critique* that I know of (which is not, in view of Heidegger's understanding of what it means to bring a text back to life, a contradiction in terms), but also because Heidegger's claim raises in the most acute way a problem that must of necessity continuously accompany any comparative study. My working hypothesis has been and will be that influence is a minimal factor in the similarities I see, and that same faith in the independent power of the matter itself would seem to speak for Heidegger's sense that a great advance can be implicit in a theory but skirted by the theorist. Moreover, it seems sensible to think that between a philosopher's study of his predecessors and his viewing of the matter itself a more and more adequate theory should come about, if only the thinker's originality and courage hold out. To be sure, unlike Hegel, Heidegger does not in other contexts believe that this is in fact the course of truth. He thinks rather that the Western tradition, beginning with a Fall, has driven itself further and further from the Being it ought to seek. For Kant alone does he make an exception: Kant's failure is not one of openness to the question but of courage to face the answer, and therefore he is Heidegger's true predecessor.

So one might ask: What about Aristotle? Might he, confronted with the mysteries stirred up by his own teachings, have come to regard the inquiry into the sources of human knowing as necessarily prior to an inquiry into their objects? Might he have moved toward a view of human thought as more productive than receptive? Certainly the beginnings of an epistemology are to be found in *On the Soul,* and there also a "making mind" (*nous poetikos*, 430 a 10 ff.) comes briefly and enigmatically on the scene. Would *he* have shrunk back from the sequel?

Yes, I think so, though not from lack of philosophical courage, but rather because of a sound sense of the slippery slope. The (pagan) ancients were over and over again on the brink of becoming modern: They were poised to discover negative numbers, steam engines, actual infinity (*against* the last: Aristotle, *Physics* 204 a 21; *for* the concept: Kant, *On the Sensible and Intelligible World,* para. 1, n. 2), representative democracy (as pointed out by Madison in *Federalist* 63), physical acceleration, the perverse will, and a whole complex of notions, probably intimately connected, that mark us as moderns. All these they were probably capable of thinking up but were more or less explicitly unwilling actually to live with. They did not so much shrink from opportunities for more advanced truth as they vigorously rejected certain siren songs of reason—Aristotle above all. Therefore, one might say, their *thought* could be made to supersede itself, but not their *thinking*: Their dogma was fair game not only to critique but also to development, but it is hard to imagine that the human thinker would have gone along, because philosophical thinking is a search rooted in a human position, an antecedent way of being. And to allow that being to be shaken in the interests of new possibilities may be an act of courage from our point of view, but it would, I imagine, have seemed like infirmity of mind from theirs. Of course, this imagined case must be conceded to be the very opposite with respect to those ancient mathematicians and scientists—the specialists—who never claimed a dedicated human ground for their knowledge and who would have been delighted with every ingenious new illumination, just as they are nowadays.

I have introduced this brief digression because it may help to highlight the persistent points of similarity between the ancient and the modern philosopher with respect to time: With all the framework changed, time is still an inner pulsing.

These are the chief passages from the *Critique of Pure Reason* that bear on time:

46–73: Transcendental Aesthetic: Time as pure sensibility
A 115–128: Transcendental Deduction, A: Phases of time
130–168: Transcendental Deduction, B: Time as inner sense
176–187: Schematism of the Understanding: Time and thought
211–212: Anticipations of Perception: Time as a continuum
218–265: Analogies of Experience: Time and space
275–294: Refutation of Idealism: Time and space

∞ ∞ ∞

A 373 ff.: Paralogisms of Pure Reason: Internality of time
438 ff.: Antinomies of Pure Reason: Time-series
454 ff.: Antinomies of Pure Reason: Infinity of time
476 ff.: Antinomies of Pure Reason: Time and cause
549 ff.: Antinomies of Pure Reason: Empty time
561 ff.: Antinomies of Pure Reason: Freedom and time
579 ff.: Antinomies of Pure Reason: Freedom and time

There are of course numerous lesser references and passages in other works that will be cited. This is the place to say that H. J. Patton's commentary on the first half of the *Critique, Kant's Metaphysic of Experience,* is of almost unfailing helpfulness.

The exposition of Kant's theory of time, interleaved with comparisons to Aristotle, will come under the following headings, which have their sources roughly in the particular part of the *Critique* indicated.

I. Receptivity and Human Finitude
II. Sensibility: Space and Time (Aesthetic)
III. Time and Thinking (Schematism)
IV. Time and Space (Analogies, Refutation of Idealism)
V. Inner Sense (B Deduction)
VI. The Phases of Time (A Deduction)

I. Receptivity and Human Finitude

The exposure of the unstated motives driving a philosophical work is a matter of empathetic conjecture, but its ultimate presuppositions are discovered in careful reading. Kant writes with perfect explicitness (though with odd marginality) about the all-inclusive concept that governs the *Critique of Pure Reason.* It is that of *representation*, the concept of all concepts, their genus (376, 74). In the Kantian context the English word is, however, an infelicitous rendering of *Vorstellung*, which means, literally, a "*fore*-position" rather than a "*re*-presentation." Kantian representations are least of all internal images re-presenting external originals; they are instead *pre*-sentations.

A representation in the Kantian sense is any modification of the soul that has the character of an object (A 99). An "object," that which is "thrown up against" something, is somehow *before* the soul, though also *of* it. The word implies that the cognitive modifications of the soul are somehow objectifying. Our soul's workings come before us, but these presentations have no interior being of their own and are never knowable as things-in-themselves, only as things for us, that is, as appearances (235). If we fantasize our soul's objects as being something on their own, we are engaged in the kind of self-delusion the *Critique* is intended to eradicate; we are exceeding the limits of our finite human reason by attributing "intellectual intuition" to ourselves.

For Kant thinks that objects can have independent interiority for us only if either we have a godlike originating intellect (72) such as is Aristotle's thought that can think itself (*On the Soul* 429 b 9)—a "self-actualizing" subject (68), or—and this is for Kant beyond consideration—if beings can come *to* us, from the outside and in their essence. But that latter case is, of course, just how Aristotle in fact conceives the human mode of knowing. For Kant beings can come to us or be known to us in their interiority neither from within nor from without (333). This condition, which begins with the limitations of our finitude, will end in the empowerments of certitude.

How can this inert ghost of an object implied in the very concept of representation gain specificity and fullness? The representational soul has, at least to begin with, two root-faculties—as was to be expected. Because if an object is to come about, some sort of psychic stage-setting is required, some duality within which, by the interplay of its powers, something can come on the scene.

Kant follows the traditional scheme: There is thinking, a higher activity, and sensing, a lower one. Thinking is done by the faculty of the understanding that is "spontaneous." Spontaneity is by no means the self-activation just mentioned. It is rather a self-production that is free enough, but in its own nature empty; the understanding functions to bring forth representations (75)—but vacuous ones. Its functions are "concepts," literally, "graspings," but grasps that by themselves capture nothing. These functions are specified by Kant into a definite number of root-concepts called, after Aristotle, categories. Later on (p. 75) rules will be given for applying these functions in an objectifying mode, but I introduce the concepts of the understanding only as it is serviceable for the exposition of time. At any rate, by these spontaneous representations the object is specified, but empty, a mere X of thought (A 104).

But we must not forget that, though we may take the *Critique* as a new human founding, its stated purpose is the grounding of knowledge. So I must cite the highest principle of all thinking that produces real knowledge: "The conditions of *possibility of experience* in general are at the same time the conditions of *possibility* of the *objects of experience*" (197). That is to say: objects and our knowledge of them are simultaneously established. So we are the makers of our world and therewith the certifiers of our real knowledge of it. This is a good place to mention that Kant's word *Erkenntnis* is not altogether captured in the flat finality of "knowledge"; it has in it hints both of effortful attainment and of recognition.

Now although the objects that constitute this knowable world are not yet fully formed, we are coming closer to time. Human finitude has two complementary aspects: It is limited in having no originating power of substantive, fulfilled thought, and it is therefore dependent on some sort of givenness. Something is "given" in the philosophical sense when it arrives as a gift, so to speak, from the outside. Such gifts are double-edged: They provide us with matter for thought; they enrich, again so to speak, the functioning but nonoriginating subject. But these gifts are also adventitious, uncontrollable, and in Kant's term, "empirical." At the outer edges of the soul some activation takes place, bringing to our otherwise impervious interior not determinable knowledge of this outside but rather an activating filler. Kant calls it sensation, and what it brings is the *matter* of empirical knowledge. Now for Aristotle matter (*hyle*) is the (relatively) featureless substrate of an actualizing form; it is that underlying no-thing of which nothing that marks Being can be predicated, and it is closely allied to the power substances have to become what they were meant to be, that is, to move toward some fulfillment, the capacity called *dynamis* or potentiality (*Metaphysics* 1029 a 3, 1042 a 26). For Kant, on the other hand, matter is itself substance, reality, presence, and a sign of existence (225, 333, 440), and his *dynamis*-analogue, possibility, is that categorical relation objects of sense have to our thinking which requires that matter be given, for without matter nothing can be represented as possible (609). That is to say, the power of becoming has gone from the substance to the subject. Aristotle's antithetical conception will prove to be at the crux of the differences their respective treatments of time do, in fact, show.

The mere object, this X introduced above, is beginning to take shape: grasped from above, filled from below. Kant takes account of this dual constitution, though not very consistently, by often using the word "object" for the empty thought and the German word *Gegenstand*, literally, "what stands against," for the sensation-filled object (e.g., 522).

Sensation is a perplexity to a philosopher who eschews at once supersensible knowledge and merely empirical knowledge. For while sensation is a necessity if reality rather than mere ideality is to describe our thinking, in its unaccountable variety—which Kant calls a "manifold"—it is an antic and contaminating basis for certain knowledge. For Aristotle there is no specific problem here. Thought, the mind, is receptive (*dektikon*) to that in external substances which constitutes their—intelligible—being, their essence; the imagination does the preparatory work of receiving and retaining the substances in their sensible aspect but without their matter, while the senses have first been actualized to begin the process. From sensing to thought there is a continuum of receptivity; thinking is regarded as "a certain kind of sensing" (*On the Soul* 427 a 20, 429 a 14).

Kant has to perform a *salto mortale*, the proverbial death-leap. He posits something unheard of, as he himself recognizes. It will ultimately be the imagination that is responsible for this novelty, and so he says of it: "That the power of imagination is a necessary ingredient of perception itself, that is something of which no psychologist seems yet to have thought" (A 120, n.). And as far as I know, he is right.

What he posits is an a priori forming receptivity, a pure inner sensibility. These two descriptive phrases are both contradictions in terms. What does it mean to be actively passive, formatively receptive? Selectively, yes; but formatively? What does it mean to sense internally, when the tradition assigns to sense the function of external receptivity? What is an a priori, pure sensibility, one not in the first instance externally affected? This was the perplexity for Cohen that was mentioned above (p. 56). If the finitude of the human constitution demanded givens, the internalization of receptivity, which must bring in its train the internalization of what is given there, reestablishes human self-sufficiency—except for that little matter of external affection called sensation. In any case, one can no longer speak of givens in the ordinary sense as something presented to us, or of sensation as being simply and directly the matter of sense. How sensation is in fact related to sense under the new dispensation will be briefly set out later (p. 86). Just how this inverted receptivity is constituted is the subject of the next section below. In any case, when Aristotle says that no one understands anything without having sensed it (*On the Soul* 432 a 4), he is formally close to but humanly far from meaning what Kant means when he says that all knowledge begins with intuition, that is, sense (730).

I should end this section with a reference to Heidegger's discerning exposition of the interrelation of receptivity, givenness, and finitude in *Kant and the Problem of Metaphysics* (paras. 5, 9, 10). But there Kant's other side is neglected, the one that allows him to write this proud phrase in his notes: "I, the proprietor of the world" (*Opus Posthumum*). It is his intimation of absolute human self-sufficiency.

II. Sensibility: Space and Time

The sensibility now to be explicated has no direct relation to the bodily senses. A possible bodily organ, a physiological common sense (*sensorium commune*), that might underlie the empirical unity of sensual representation as the sensibility does its pure unity, could never be the seat or organ of the soul itself. For that would mean that the transcendental ground of all appearance would itself appear. Insofar as the soul is among bodies it has "*virtual* presence." But to seek it with and through the body is to seek the square root of two: it is to attempt the irrational (*On the Organ of the Soul*, 1796).

The sensibility of the *Critique* is, as any sensibility, first of all a capacity for being affected. But it is, in distinction from our bodily sensing, "transcendental." "Transcendental" is a fourth word in the interrelated vocabulary of the *Critique*: critical, pure, a priori (25). All four mean, from different perspectives, the same thing: having to do with the nonempirical, presensory capacities that first make experience possible. The transcendental faculties "go beyond," or come before, the objects they are responsible for (and in being acknowledged prevent us from *transcending* our knowledge-warrant).

The study of this pure capacity is therefore called the "Transcendental Aesthetic"; aesthetic is what has to do with *aesthesis*, the Greek word for sense (35 n.). Kant's study is conducted as an analysis, a reduction to elements through abstraction, that is, through thinking away the empirical impurities on the one hand, and through the isolation of the capacity for sensing from thinking on the other (36, 46). These practices do not, however, yield the sensibility as empirically derived—they merely allow it to appear. Thus will be found the conditions sought, those that will make possible an experimental knowledge that is binding. In the analysis, the sensibility, the soul's organ of sensation—Aristotle too speaks of the receptive capacity (*nous*) as the soul's organ (*Problems* 955 b 25)—comes earlier in the approach to the human constitution than thinking, since thinking is higher and more integrally related to, in fact *is,* the subject. But which is ultimately deeper, sense or thought—that remains a never-ending perplexity for readers of Kant (p. 83).

The governing word of the Transcendental Aesthetic is intuition. "Intuition" is the translation of *Anschauung*. The German preposition *an* means neither an intensive nor a directional "in" (as in *in*flammable or *in*sight) but rather "at." The intuitional capacity is for viewing. But Kant uses the word in crucial places also to mean something like the work-product of the sensibility, its content: a pure intuition (160). So intuition is both the form and its pure content—a complication, but an uncircumventable one, which will be taken up just below (p. 64).

What characterizes the receptive sensibility's intuition is immediacy and particularity. No discursive, that is, process-requiring, thought intervenes between the sensing and its given objects, and all its objects are particular presentations, not abstracted generalities, that is, not gatherings of thinking, concepts. To sense is to have *this* given directly before one. What is novel is that such a given will emphatically not be a "this something" (*tode ti*) in Aristotle's substantial sense.

The affective character of the receptive sensibility is now to be further specified. We get to glimpse it by thinking away the manifold of happenstance matter so as to be left with the intuition as content and as form. Now the indeterminate, that is, unspecific, object of sense is called "appearance," whose matter is sensation (34). So it is the sensory appearances that we must wipe away from our sensibility to get pure intuition in both senses.

Kant finds *two* "springs of knowledge" (55), two forms of sensibility, space and time. He considers other springs as thinkable, though not knowable, but these are the ones he *finds* (36). It just is part of our nature to have a dual sensibility that expects to be affected (75). So a first question, Why these two, space and time?, has no answer in principle. It is how we are. Aristotle would not agree: the common sense, his sensibility-analogue, is primarily receptive of motion, and space and time are sensed only derivatively (*On the Soul* 425 a 18). Motion for Kant requires sensation and is empirical (58), and so it is excluded from pure intuition; the mobile is not primary. At any rate, no justification is given for specifying the sensibility into just space and time. Not even in Kant's dissertation of 1770, *On the Form and Principles of the Sensible and Intellectual World*, announcing what will become the transcendental sensibility, is such a justification forthcoming. The motive, however, is not far to seek. Space and time are, after all, together with mass, the primitive dimensions of Newtonian dynamics, the science for which the *Critique* is meant to provide the grounds of possibility.

But a second and third question do have—related—answers of a sort. Why are space and time expounded as parallel and coordinate? Why are they nonetheless presented in this particular order? The immediate answer is that insofar as they are both intuitions they can be presented in formally identical expositions, though if we go deeper into the *Critique*, their relation both with respect to their nature and their priority becomes more and more questionable. In *On the . . . Sensible and Intellectual World,* time is presented first, so it is plausible that Kant reversed the order for a reason, probably because the presentation in the *Critique* is from the intuitively representable outskirts to the inferred hidden center of the human cognitive constitution. As I said, the further we go, the more it becomes a question where this center really lies, in space, time, or thinking. That question will be taken up in section IV, Time and Space.

1. *Space*

Space is the "form" of outer sense. Form is conceived by Kant in the Aristotelian way, as having its complement in matter. The matter of which space is a form was, as we saw, sensation. Sensation is multiplex and variable. It is what Kant calls a manifold, an unshaped givenness in need of formative ordering relations. Since sensation can hardly be ordered by sensation, there must be a prior, an a priori form within us, Kant argues (34). But this merely critical requirement turns out to have a strong character that it imposes on the sensation it receives. Never was there a more imperial receptivity.

Space, then, is the form of "outer" sense. "Outer" is a fascinatingly equivocal word in the *Critique* (38). It means "outside us," that is, "distinct from and appearing to" consciousness; it is *pre*sentation in that most vivid sense in which spatial contents do come before us as external—though of course "before us" still means "within us." But "outer" also means "in a spatial mode"; spatial representations are characterized by having their parts outside each other and simultaneously present, that is to say, by being extended. A third meaning of "outer" is only implicit: Kant sometimes speaks, as I have done, of "in us" as meaning within the soul (e.g., A 373); "outside" would then mean in the inaccessible supersensible field beyond our ken.

We may say, then, that the outer sense is a sense both of externality and of extendedness. It is also fair to say that Kant does not attend to the equivocation that, were we speaking of ourselves as embodied, would be easy to resolve: The spatial world is outside us as bodies by the same rationale by which its parts are outside each other. But in the little communication on the organ of the soul cited above, Kant makes it clear that this solution is not to be thought of. The problem is really deeply embedded in the representational character of human knowledge. Our outer representations have one sort of distance as being beside each other and another as being before us and consciousness. This problem, I want to say with confidence, Aristotle would have seen at once as a difficulty arising from internalization of space or place.

If then we think away all sensation-filled appearances from space, we discover that space itself remains and is rich in determinations, the sort that yield the pure a priori synthetic judgments belonging to a full-fledged, unconditionally certain, pure science—geometry (64). But first we most notice that vestigial space is in fact a sensible given according to the criteria: It is both represented immediately, without any intervening conceptualizing, and it is one single particular whose parts are not different spaces but just that—parts of the same whole. Then come to view those determinate relations that could never be contributed by sensation, the geometric intuitions in the ordinary sense. They range from deep topological features such as the fact that certain figures, like our left and right hand, are, in modern terms, "incongruent counterparts," figures whose nonspatial, conceptual description is identical, but that can yet never be oriented in such a way as to cover each other in space (*Prolegomena*, para. 13), to elementary Euclidean facts like the angle sum of any triangle whose equivalence to two right angles can never be derived from the mere concept of triangularity.

The reason I cite Kant's examples is that they reinforce the necessity for introducing, as Kant indeed does, a middle entity between formative intuition and conceptual thinking—the aforementioned content, the true bearer of the particularity of space. In it the geometric figures are "exhibited" in accordance with their concept (741), that is to say, the pure imagination inscribes them in the pure field as thinking dictates and the field prescribes.—The psychological counterpart of this transcendental procedure is perfectly familiar: We think "triangle" and then draw one into our imaginative field. More will be said of this in the next section.

What is of interest here is that the intuitive content found in space is not so easily made evident in time. In fact, Kant repeatedly says that time has no independent science of its own (*Metaphysical Foundations of Natural Science*, 471; *Critique of Judgment,* First Intro., X). This is clearly because it does not have the extended territory of space. And so he makes no effort to sketch out a pure temporal intuition-content. Herein lies the root of the increasing asymmetry between time and space. Nonetheless, time hosts something analogous to geometric inscription: *counting*. And this, I think, will bring Kant close to Aristotle (p. 73).

In the detail of the spatial exposition we must never lose sight of this strange new human world—all within us yet external; thoroughly formative yet receptive; absolutely pure, yet filled with given content.

2. *Time*

It seems more natural to see time located within, since only the crudest naturalists think that time minus consciousness is simply out there. And yet, if space has its difficulties, time will have them raised into enigmas—old enigmas in a new framework.

Kant makes short shrift of the "inner" sense at this point (46–49), though it will appear over and over in the *Critique*. What "inner" really means is left for section V, Inner Sense. The exposition itself is fairly parallel to space. Metaphysically speaking, time is not empirical, but a priori. For if it were empirical we would not, as we do, perceive simultaneity and succession as necessarily universal. Of course, for Aristotle these two are inevitable properties of every motion, and in that sense empirically universal. Appearances, says Kant, can be removed from time but not the reverse, which shows that time is cognitively prior. Yet time is not a concept but a unique particular, because all times belong to it as parts. Hence time is a pure intuition. Transcendentally viewed, a priori time makes possible a number of axioms necessary to the empirical science of motion, because it is unidimensional and progresses by before-and-after.

In fact, change itself depends on the a priori intuition of time. Motion, that is, change in space, has time as the condition of its possibility. For motion is formally the connection of contradictory determinations in one existing thing (291); for example, a real thing is moving if it is now here, and now not here. This is a breach of the logical law of contradiction, one which is made possible only by time, insofar as the law need not hold for *different* times, only for one and the same time. Indeed, the law itself might be said to depend on time, because it holds only when qualified by a time-feature, namely simultaneity: A thing cannot be both A and not-A at the *same* time (*On the . . . Sensible and Intellectual World*, para. 14, 5). So successive Becoming and present Being, to use ancient terms, are equally time-dependent for Kant. Time allows logic to become fluid and existence to be determinate (48). Aristotle would object here that both Being and Becoming are in the mobile itself; but more of motion in section IV, Time and Space.

In the time section Kant criticizes certain powerful current rival theories, especially, though anonymously, that of Newton, who posits time as an absolute, real, equable flux (*Principia*, Scholium to Definitions). This notion, Kant says, produces an *Unding*, an "un-thing," a self-canceling concept (349). For such a flux is an eternal, infinite but vacuous reality that is there in order to contain within itself all that is real (56). This monster-notion is attributed by Kant to the party of natural scientists in general. And indeed, I can think of no philosopher who has held it, though physicists have eschewed it as well (pp. 14, 153).

Kant does not give an explicit critique of Locke's insouciantly radical theory of time, perhaps because he has already set him—respectfully—aside as a "*physiologist* of human understanding" (A ix). Locke is the only author I have come across who cites Augustine's perplexity about the elusiveness of time (p. 117) with irony, being confident that a plain answer can be given: We derive our idea of time "from reflection on the train of our ideas." Duration, in turn, is furnished to us by reflecting on the "distance" between any parts of the succession of ideas "or between the appearance of any two ideas in our minds" (*An Essay Concerning Human Understanding*, 1690, II, xiv, 2–3). By "ideas" Locke means mental representations either of sensations or of the operations of the mind itself. Locke's derivation of duration seems to me to bring in by the back door the Aristotelian problem of pure psychic motion (p. 46), for how else can the mind measure the distance *between* any two ideas? Although Locke's solution is, for all its radical plainness, embroiled in old problems, he can be credited with being the remote originator of two notions that come to the fore only two centuries later: the stream of consciousness, that is, "the train of ideas," and the constitution of time from inner phenomena, that is, the successive appearance of ideas and intervals (p. 208; Gallagher, 8). Kant might have used this occasion to specify his critique of a thinker whom he rejects as an empiricist but appreciates as the first epistemologist. But to return to Kant himself.

In sum, then, time and space are the subjective conditions of knowledge and not objective features of things (49). But this does not mean that appearances in time and space are illusions. They are precisely *appearances*, perfectly real for us when filled with sensation. Kant's frequent formulation is to the effect that these two forms of sensibility are transcendentally ideal as capabilities of the soul, and empirically real when receptive of given matter (70).

III. Time and Thinking

Intuition on its own is blind. In fact, there is no mere intuition: we "look at" nothing in advance of thought. The pure intuition, divorced from sensation at one end and from understanding at the other, is only an analytic residue. All human awareness is thought-structured a priori, from the beginning. Thus to intuit even a pure triangle, for example, I have to *think* the concept of a figure enclosed by three lines, and *I* simultaneously have to think it: a determinate concept and its unifying subject.

1. Concepts

The understanding functions through concepts as the sensibility was affected through intuitions. As has been said, Kant borrows the fundamental concepts from traditional Aristotelian logic (107) and selects them with a view to the shaping of an object-representation, the transcendental *thing* that sensation will fill to produce an empirical *body*. Such a body is not a self-moving independent natural substance, as is an Aristotelian nature; instead its structure is pre-given in us and only its matter is "real," as we will see. The Transcendental Logic, the rational counterpart of the sensible Aesthetic of the last section, analyzes the understanding into its elementary concept-functions. The two capacities, the "aesthetic" and the "logical" one, are, at least to begin with, *utterly* different. To get the full force of the development, we must keep that point well before us.

A category is only a function, not a real object in the sense of *Gegenstand*. For Aristotle abstraction left a residue of thought without Being. For Kant there is a kind of originary abstraction, one that comes before the object is even on the scene. These abstract empty functions that make up our thinking faculty are the categories. They will be brought in below as needed.

The original activity of all thinking is unification: "synthesis" in Greek, "composition" in Latinate English (103). Such compositional unification is the task assigned to the understanding in light of the manifold, the material variety, it will be asked to master in the receptive sensibility. The understanding is, accordingly, not receptive but spontaneous: it brings forth its own representations, though they are mere empty functions. As the understanding effects synthetic unification in the lower faculty, so it also reaches upward (or deeper) into our constitution: It specifies, or enacts, or even *is* the unity of the underlying subject, the "I think" behind all our conceptual acts (132 ff.). It is a vexed question just how the functions of the understanding are related to the self-consciousness that is captured in the word "I," a question set aside here. However, the transcendental subject will make its appearance in time, and that will be the occasion for saying more of the "I" (p. 79).

2. *Judgment and Imagination*

The two abysmally diverse faculties of sense and thought *have* to cooperate to make us the arena for the appearance of real objects capable of being compellingly known. Kant keeps the section of the *Critique* devoted to accomplishing this very brief (176–87), as is his wont when he comes to the crux of his projects. We are here at the intellectual center of the human cognitive constitution.

One of several different ways to pattern the upper faculty in the *Critique* is tripartite: understanding, judgment, reason (169). Concepts of understanding subsume particulars; reason makes inferences; and in between these, the judgments of the judging power (both are called "judgment" in English) connects terms not, of course, into the audible propositions of logical speech, but into transcendental a priori judgments.

Judging is the real transcendental work. It connects concepts to appearances in consciousness according to rules (171), and so it gives the conditions of possibility of making real judgments in life. One way to see the difference between a merely logical judgment and the transcendental judgments that make knowledge possible is this: The law of contradiction governs all logical judgments, for a thing can't be A and not-A. But it does not hold unrestrictedly for the crucial kind of judgment we are talking about, the transcendental kind, and that is, of course, because time comes into play (48, 192; p. 65).

Recall once more that categories have no use except as applied to the appearances in intuition (p. 59). And these are only obscurely preformed until the understanding determines them (304, 185). To the judgment, then, is assigned the crucial bridging function between the two root-capacities. We do in fact judge that "the plate is round," but it is a judgment easier to pronounce than to explain as an accomplishment of the soul.

For it is a primary reflective principle accepted by Kant that in every judgment the representations connected must be homogeneous (685). Thinking is by itself closed to connection with anything but thought. Similarly for sense. And this isolation, the result of constituting the human being from receptivity and spontaneity separately, must now be bridged.

Kant introduces a deep new power, the *transcendental imagination*. It produces its works by a "hidden art in the depth of the human soul" (181). More will be said of this power itself below (p. 87); here I will concentrate on its product, the schema. The position of the imagination in the design of the *Critique* can be discerned thus:

Judgment
/ \
Intuition Concepts
\ /
Imagination

It is a faculty that works underground to homogenize thought and sense, so that judgment may connect them in the light of day.

Imagination in the ordinary cognitive meaning is the ability to see an object without its presence (151). This understanding descends straight from Aristotle (*On the Soul* 425 b 25). So also does that mediating position of this power between intellect and sense, which brings it in one respect close to thought and in another close to sensation (428 b 13, 427 b 10). And so also perhaps does the reference to its obscurity (414 b 17). But here is a juncture where the ancient and the modern philosophers become well and truly incomparable: The Kantian imagination has by far the deeper work to do.

3. *The Schema*

This work is the transcendental schema, a "third," mediating between category and appearance (177). What the schema must bring about is plain enough. It must bridge the distance between thinking and sensing so that the categories can get their grasp around appearance and intuition can be thought-imbued. The result will be both the constitution of the object of real knowledge and the grounding of the possibility of that knowledge itself.

That is what the schema must do, but how? It might be said here that in any theory of knowledge thought and sense must be brought together merely because they are in fact together. Kant too begins with this human fact: There *is* knowledge, pure and empirical, to be found in mathematics and physics (B x ff.). Kant is in a better position than most because for him both are internal, and so in one respect the two faculties are already on the same ground. If in other theories it is a plain mystery how thought grasps space or the soul marks time, in the *Critique* it is a technical perplexity—though perhaps one that masks the deeper enigmas.

The actual technical problem is whether the schema that Kant sets out is more like thought or more like sense. He gives hints either way—or perhaps he is speaking of two different types, a rule-governed type for time and an image-governed type for space (Makreel, 30).

This much *is* clear. The schema is not an external picture, such as the dots representing numbers as they are counted up (15), not even a mental image in the ordinary sense (181), though it is what makes such empirical imaging possible.

Positively speaking, the schema is called a rule, a method (179–80) for representing, say, a quantity, in a picture in accordance with a concept. Sometimes the emphasis is more on the quasi-pure spatial product itself, a generalized picture, such as a specieslike sketch of a dog. Kant calls it a "monogram," a "hovering" drawing, as it were, "not determined by any assignable rule" (181, 598). The true schema is produced unambiguously by the power of imagination (*Einbildungskraft*, literally, "the power of inward picturing," 180), but it belongs, more ambiguously, to the understanding. Thus Kant speaks of the schema of the concept of the understanding and of "the procedure of the understanding with these schemata." The very title of the section, "Of the Schematism *of* the Pure Concepts of the Understanding," is ambiguous. It isn't quite clear whether the *of* italicized by me is a so-called objective or subjective genitive, that is, whether the understanding is the object of the schematization procedure or the performer of it. In any case, a transcendental imagination that is a faculty of rules and an understanding that permits schematization are two faculties assimilated to the point of identity, and this circumstance will, of course, underlie Cohen's categorization of time and Heidegger's temporalization of the understanding. To me it seems inevitable that the two powers should flip into each other this way. The judgment of the understanding as a faculty of subsumption is nothing but the potentially conscious counterpart of the ever-hidden transcendental imagination as a bridging power. It is the one resolution of the sense-thought mystery possible—that they should be two streams from one source. But it only works if the knowable world is internal, and if the human finitude in question is not that of a being among beings or a creature among creatures but that of a self-enclosed subject with an aboriginally dual cognitive constitution. I will be opting for Aristotle's natural mysteries in Part Two of this book.

The mission of the schema in general is then to provide the condition under which the categories can be applied to objects and attain significance (185). Because of the equivocation just noted, we may speak of this application in two ways. (1) The concept or category is schematized, sensibilized, sensualized, temporalized (or spatialized), time-determined, time-affected. Or, the other way around: (2) Time is categorized, thought-determined (186), thought-restricted.

4. *The Priority of Time*

There is good reason to speak in the second way, at least in the present context: It is time that is being thought-imbued. Why time rather than space? Kant makes it clear that the schematism applies equally to space (181); geometry is the prime example of thought determining space. Moreover, as we shall see, time has no pure counterpart-science to show (p. 72). Yet it is time and time only that Kant treats of in the section on the schematism proper (182–85).

Here we see time becoming progressively primary in the *Critique*—until space suddenly regains ultimate primacy (p. 78). There are two opposed yet complementary reasons for the primacy of time at this point. As inner sense, time is, from the subject's point of view, the place of its own appearance to itself and so the more intimate sense (p. 79). But it is also the more inclusive concept (*Inbegriff*, 220), the one that contains *all* our representations.

5. *The Schemata*

So far I have spoken of the schema as a general notion. But there are different schemata for temporalizing each category. Together they define what determinate time is, as distinct from the pure form of intuition and its blind intuition-content.

So half the answer to the question "What is time for Kant?" is to be found in these details. The other half returns to the pure form and its peculiar content as the inner, now meaning "inward-looking," sense (p. 79).

There are four groups of categories (106) and the schematism follows these. They give in turn the time-series, the time-context, the time-order, and the inclusive concept of time with respect to all objects (184). Together these time-determinations, which are about to be elucidated, make up all our conceptions of time. Kant is brief to the point of abruptness about the "dry and boring dissection" of the schemata (181)—the one point in the *Critique* where this man, whose name is probity, sounds plain implausible.

(a) The first category—Kant uses only one of the categories in this and the next set—is the category of quantity (Aristotle's *poson*): The pure schema of *quantity* in time, Kant says, is *number*, where "number" is the only schema-name italicized. He continues:

> Thus number is nothing else but the unity of the synthesis of a manifold in a homogeneous intuition in general [brought about] through this, that *I generate time itself* in the apprehension of the intuition. (182, my italics)

Since this schema is crucial to my point, which is that for both Kant and Aristotle the generation of time is in counting, subsection (6) below will be devoted to number and numbering and its relation to inner sense. Let me say here that just as quantitative thinking in space determines spatial continua such as lines, so in time it brings forth discrete elements, monotonous iterations, namely the time-series that unifies the variety of sensation as always different but always equally present: now_1, now_2, . . . , now_n. Just as Aristotle thinks of the Now as providing continual (though *not* continuous) contact with sensible presence, so does Kant; this transcendental moment is called "apprehension." As producing the perceived present it will be treated under section VI, on the phases of time.

(b) The category of reality is the pure concept that is fitted for the conceptualization of the real in sensation. When it determines time, the resulting schema calls for time to be filled with a continuously waxing and waning degree of sensation. This schema allows the empty object of thought to be a real *Gegenstand* (148), an object that has matter and, one might say, that matters, has pertinence (*Sachheit*, 182). Thus time gains a sensory, real content but also the ability to be empty. For Aristotle the apprehension of time empty of motion ought to have been a real problem, if the counting of motion *is* time. For Kant this problem does not arise, since the counting that generates the time-series can go on in pure intuition even if the sensation-content goes to zero.

(c) The third general category-group is that of relation, and all its three categories are schematized; the relations are substance and accident, cause and effect, reciprocity of action and reaction.

The schema of a substance grounds permanence in time. The real in time given by the last schema is rescued from the flux of the time-sense and given the possibility of permanence. What one might call "substantized" time yields the notion of a substrate in and through time. Such a substance makes possible an "empirical time-determination in general" (183), which I take to mean time-measurement. For time itself does not elapse, but the existence of what is changeable does. Unchangeable time therefore needs a correlate in appearance against whose background (or underground) changing appearances in their simultaneity and succession may be determined. This correlate is the "constant and enduring appearing substance of things" (*substantia phaenomenon*, 186). The implications of this schema are not innocent, as will be seen in the next section (IV). Recall that Aristotle has an analogous requirement of physical permanence; it is the individual moving substance. But what Kant calls substance, namely pure substrate (231), is what Aristotle would have called matter (as in fact Kant himself does, 321), just as what Kant calls reality in the previous schema, namely sensory qualification, is what Aristotle would have called accident. In the two schemata about to follow, Aristotle could have recognized nothing, for these are the modern center of the *Critique*: the conditions of possibility for the laws of classical physics.

The schema of cause and effect in thought determines a necessary succession in time: "Causalized" time appears as a necessary order of real events. Hence time appears in reality as a necessary forward progress, with a determinate before-and-after. Thus the Aristotelian "before and after" that followed from the development of a natural body, that is, its actualization, and whose stages are not cognitively determined but follow from its nature, has migrated into the sense of time itself: Not "time is the number of motion with respect to its before-and-after," but "motion is the real appearance of the before-and-after belonging to causalized time."

The schema of reciprocity, finally, makes it possible for time to contain an ordered simultaneity. When time is conceptualized in terms of a reciprocal relation, it appears as the system of simultaneous causality, interrelated nature at any moment.

Note, incidentally, that these five schemata build on each other: Mere succession (counted nows) is filled with variable reality (accidental sensation) and its permanent substrate (substance-in-appearance) against which necessary succession (cause and effect) can now appear and therefore also reciprocal interaction (simultaneous mutual causation).

Let me recall for a moment the stated aim of the *Critique*, which is finding the grounds of possibility of a necessary and universal science of nature. In the last three schemata the transcendental preformation of three natural laws of classical physics is to be recognized: In the schema of permanence, the law of the conservation of matter; in the schema of cause, Newton's first law of motion, the law of inertia, which says that there is no change of motion without a causing force; in the schema of reciprocity, Newton's third law of motion, which says that to every action there is always an opposed and equal reaction (Newton, *Principia*, "Axioms or Laws of Motion"; compare in the *Critique*, the three Analogies of Experience, 224 ff., and in the *Metaphysical Foundations of Natural Science*, "Metaphysical Foundations of Mechanics," 541 ff.).

The last set of categories to be schematized has to do not with the features of time itself but with the logical "mode" in which appearances in time are to be thought: as possible, actual, or necessary. A thing is possible if it agrees with the temporalized law of contradiction, which says that contradictory properties cannot belong to a thing simultaneously but only in succession. A thing is actual if it exists at a definite time (265). And it is necessary if it existed at all times. To Aristotle these relations would have seemed partly familiar, partly upside down. The law of contradiction is stated by him with exactly Kant's time qualification and is for him, as for Kant, a law of Being before it is a rule of rational speech (*Metaphysics* 1005 b 23 ff.). Moreover development is understood by him as a becoming, a successive going from not-being something to being it (1089 a 28). Also like Kant, Aristotle recognizes a certain motion as existing at all times, namely sempiternal heavenly motion (1072 b 11 ff.). But the notion that actuality should be existence, that is, being here and now, would have seemed to him false, for when a being becomes actual, when its capacities are fully developed and at work, it is in that respect atemporal.

6. *Counting*

> Thus number is nothing else but the unity of the synthesis of the manifold . . . through this, that I generate time itself. (182)

Here Kant says elliptically but plainly that number comes about as I generate time. To puzzle out this sentence seems to me of prime importance in understanding what time really is for Kant.

The first thing is to note yet another asymmetry between time and space: Space has a pure science of its own, geometry, but no such pure science exists for time. To be sure, as the generation of figures in space takes time (203, 283), so the representation of time requires space (292). But that is just where the asymmetry shows up: It takes time incidentally to inscribe a figure according to a concept, but it requires space inherently to get a representation of time, be it of its continuously flowing or its discretely pulsing aspect (as will be shown in the next section, IV). Here is a way to put it: Time is inherently imageless, and its fieldless unidimensional representation, the fact that its figure is a mere line, is not really the cause—for how could a representation be a cause?—but merely the symbol of the imagelessness of time itself. It is interesting that counting, the best candidate for appearing time, is ill-matched with space: In the *Critique of Judgment* (para. 26) Kant points out that measurement, the application of number to space, is always merely subjective because (Euclidean) space has no absolute unit.

Three possible candidates for a science specific to time come to mind: (a) motion, (b) arithmetic, and (c) psychology.

(a) Motion is indeed said by Kant to be the arena of time. Its science, however, is not pure but rather empirical (58, *On the . . . Sensible and Intelligible World*, para. 12). Consequently, the proto-laws of motion spoken of above do not become a system of scientific truth until the empirical element of *real* change is added, which requires matter and bodies.

(b) Arithmetic *is* a pure science, though not a temporal one. It became a candidate for the science of time only because the sentence to be explained mentions number. There are many passages from the *Critique* and from Kant's letters bearing on the atemporality of arithmetic and its nonaxiomatic character (Friedman 1992, ch. 2).

Number, however, as the object of arithmetic is not number as generated in counting. Time may not own the science of number, but it does make that science possible by generating number: now_1, now_2, . . . , now_n. Several passages from Kant's correspondence and later works show, in words similar to those quoted from the first schema, that number is transcendentally constructed through the synthesis of the imagination by the successive addition of units (Friedman 1992, 115 ff.). The best passage comes from the *Critique* itself; it speaks of the successive positing of units based on time that makes magnitude in general possible (300).

So while time does not yield a science of number it does underlie the counting "without which no object of mathematics can be given" (Friedman 1992, 115). The transcendental procedure, the schema of number, thus makes possible "progressive iteration," or in Kant's terms, "successive repetition" (300). To put it vividly: When the concept of quantity determines time, time begins to beat or pulse. These units are to be thought of as contiguous; there is no time between them any more than there is a "space" between two counting numbers. The Now with a subscript is, so to speak, that pulse, and in that sense I generate *appearing* time within myself: It now appears as an inward pulse that monotonously beats out a number, and the frequent references to the homogeneity attached to the schema of quantity make sure that the units of this number are *mere* units (182, 162). How the beating out of units can actually result in a *numbered* conclusion is explained by Kant under the heading of "synthesis of reproduction" (A 102), which will be explicated below (p. 88). Meanwhile Kant makes it quite clear that the generating of a magnitude and of time, moment by moment, is convertible; he speaks of the part-by-part generation of a time and says that it is just so with time-bits: the successive progress from moment to moment generates a *temporal magnitude* (203).

We see, then, that although for Aristotle motion is the substrate of time while for Kant, conversely, time is the ground of motion, both are driven to the notion of some sort of psychic counting. For Kant, such original counting *is* thought-quantized time.

(c) So we come to psychology, the study of the soul, which belongs to time, because time is the inner sense, the sense of oneself (p. 79). Kant repeatedly denigrates this study in both its pure and its empirical form. He considers the latter poor in insights because it cannot, being of a one-dimensional object, be properly mathematicized, except perhaps in respect to its law of continuity (e.g., *Metaphysical Foundations of Natural Science,* Preface, 471). This very restricted view of our ability to know ourselves as temporal beings seems to me to be exemplified in Kant's low opinion of music. He presents this most purely temporal of arts as a mere pleasing play of sensation and affects; this opinion is expressed throughout the *Critique of Judgment,* which deals with aesthetics in the modern sense as the study of taste (e.g., para. 51), and even in his otherwise wonderfully observant *Anthropology* (1798, para. 18). Leibniz, in the *Principles of Nature and Grace* (1714), takes just the opposite tack: Music is the unconscious counting of the soul and a pleasure given to us by God so that we may know of him (para. 17); in music the soul is revealed to itself and God to it. For Kant the transcendental temporality of the subject overbears the appearing time of the human soul.

Rational, that is, pure transcendental psychology is of even less standing. The center of the *Critique of Pure Reason* is, after all, devoted to the critique of any attempt to know the soul by pure reason; the endeavor is transcendental illogic (390 ff.). Our "I think" is inaccessible to mere thought. What we do know of it we know only in temporal intuition, and it is not much (p. 83).

Kant differs, *toto caelo*, from Aristotle in this denial of rational self-knowledge; thought thinking itself is the very high point of all knowledge for the latter (*Metaphysics* 1072 b 21 ff.). But they coincide in a point crucial to my argument. Just as Aristotle, almost by the way or even inadvertently, refers to a certain motion of the soul, so does Kant: He speaks of "motion as an act of the subject (not as a determination of the object)" and gives as an example of this sort of motion the inscription in space of geometric figures, which is "a pure act of successive synthesis" (154 and note). So any pure act of the soul that takes time—and all do—implies some sort of inner motion. This consideration raises the fairly devastating suspicion that transcendental time after all either is or implies a motion of the very soul itself, even for Kant. Husserl will articulate this dilemma (p. 149).

Kant's inattention to the non-naturalistic study of the soul may be a little disappointing but it is not surprising. For him the human being truly knows itself in two other primary ways: as a person through the sense of self-respect that comes from employing the rational will consistently, and as a knower in contemplating the image of its own cognitive constitution in sensation-filled space, in natural science. This last point must now be shown to hold within Kant's system.

IV. Time and Space

The deeper we progress into the *Critique*, the more space comes to the fore. The driving question, the more pressing for not being stated explicitly as a problem, is: However, in the absence of a proper representational field of its own, can time appear? And the answer is: In all but one respect, *only in space*; the excluded respect is the appearance of the representation "I think" in the inner sense (section V). Kant reiterates this revelation especially in two places, in the "Analogies of Experience" (224 ff.) and in the "Refutation of Idealism" added in the second edition (274 ff.) and expanded and amended in a long note to the preface of that edition (xxxix). So there can be no question about the importance of this insight.

If the schematism of the understanding gave, through the schemata, the rules by which concepts apply to appearances, the following section details the fundamental rules, the principles, by which the understanding actualizes the schemata so as to obtain a system of judgments yielding experiential knowledge. One might say that the schematism of the understanding explains the possibility of the objects of nature, while the principles of the judgment explain the possibility of the science of nature, in accordance with the highest principle of all synthetic judgments. That principle, recall, says that the conditions that make experiential knowledge possible are identical with those that make its objects possible (197). We know what we make.

I might add here that the term "rule" is easier said than grasped. The understanding is our "faculty for rules" (A 126), and in the present context a rule is a condition for unifying the manifold of appearance into the thought-structured composite Kant calls a synthesis (e.g., 103). In a word, a rule is just the articulated expression of the understanding at work. And its work, however accomplished, is: *E pluribus unum.*

1. *Analogies of Experience*

Most important for my purpose is the set of principles called the "Analogies of Experience." It is remarkable that the preceding set, the "Axioms of Intuition," is not at all occupied with temporal number, for these Axioms correspond to the first schema in which numbering was said to generate time (182). The reason given is the lack of axioms of arithmetic compared to geometry (205; p. 64): Consequently only the application of geometry to appearance is guaranteed in the axioms. Thus nature is primarily geometrically determined, extended space, and number is primarily left to the discretely thinking soul.

The Analogies have this name because these rules allow us to conduct an analogous inquiry: to look for empirical unifications just as we established transcendental ones. As the schema required substance that was permanent, so the experiential science of nature allows us to assume a matter and to search for its quantitative persistence (223). Although I will bypass here the science-grounding aspect of the principles, it is clear that we are getting down to brass tacks: from conceptual to schematized to experiential judgments. Moreover the Analogies are especially concerned with existence in time and space (220).

That is why the features of time as it is related to space become a special focus of this set of principles. These features are already familiar: permanence, succession, simultaneity (219). But now we will learn how they come to appear.

(a) The first principle is that of the permanence of substance (224). We are familiar with it from the corresponding schema (p. 70) and the example above, except that now a reference to perception, that is, to empirical, sensation-filled consciousness, is added: "In the object of perception, that is, in the appearances, a substrate must be met with which represents time in general" (225). The reason is that time in itself is not perceptible. And it is not perceptible because it itself is not a sense-affecting appearance but the singular form of all appearances, the receptive condition of their being given to us. As such it does not itself change—recall that Aristotle says this too, but only because time is a mere property of change—but rather changes, succession, are represented as determinations of time.

What is the problem here? Why does time need substance, "the substrate of all reality," which is the bearer of sensation and of all existence, in order to become perceptible? This problem has been raised above (p. 71), but here is yet another way to consider it: Space is not only the *form* of intuition, it is also *an* intuition (160; p. 62), and thus has intuitional content, which, when it is sensation-affected, becomes perceptible. Not so with time. As an intuition it is elusive, though in the next section (V) another attempt will be made to tease out its rather thin features.

That thinness drives Kant into bringing on substance. Philosophers and laymen have often entertained the presupposition that something substantial underlies change (227). But it has never taken on just this guise, that of a mirror of time, or better, a representation of time, *both* in the ordinary sense of standing for it, and in the Kantian technical sense of being a modification of the soul (p. 58). Kant says that permanence of substance "expresses" time as the steady correlate of all appearances in their existence (226). We will meet this "correlate" again (p. 81).

There follows a radical revision of the concept of change. For Aristotle it was the development from capacity to fulfillment of *a* substance or *a* nature. Kant now amends this or any derivative notion: Change is nothing but a way of existing that follows upon another way of existing of the same object (*Gegenstand*, 230)—it is again that law of contradiction made fluid. And substance has no other power or end than that of holding still under the flux of its accidents. Substance is nothing but a supine substrate, mere matter—*hyle* is Aristotle's term—which becomes, with slight additional specifications, the movable matter of physics (*Metaphysical Foundations of Natural Science,* 542). Thus the source of change is elsewhere, not in nature. In law-governed change it is in the cognitive subject, and for adventitious variation in an unknown outside.

(b) The Second Analogy is that of the law connecting cause and effect (232). It turns out that the change that was mere succession of being and nonbeing in the determination of substance provided only for duration (183) but not for the driving forward-order of time that gives it before-and-after. In mere thought or in my imagination I may subjectively connect two situations in any order, but in perception I am by this principle entitled to assume an objectively necessary succession (233). For example, if I sense warmth in a room I may look for the oven that was hot first.

The same principle that orders time successively through indicating the causal connection of appearances even apart from our perception also guarantees that it appears as continuous. This fact is only intimated in the "pure" *Critique* (244, 256); it depends on the conversion of matter into force (321). This conversion, in turn, depends on the identification, accomplished on a more empirical plane, of substance and causality; it is made in the *Metaphysical Foundations* (e.g., 524). So time acts causally as it appears in substance: steadily, continuously. Since Kant himself never explicitly establishes the continuity of time, as he does its specifically temporal modes of succession and simultaneity, the foregoing is speculation—though he himself says that the continuity of time can be known only empirically from appearances. And he says it in italics (244). Evidently time is *not* representable in pure inner intuition as a *continuum*. That, of course, goes very much to my point: Time is *discrete* inner counting, as for Aristotle.

(c) The Third Analogy says that all substances, insofar as they are perceived in space as simultaneous, stand in thoroughgoing reciprocity of action (256). This principle needs two elucidations to make the sense we expect. First, Kant is not strictly entitled to the plural "substances," but again the *Foundations* help. He must mean the transcendental prototype of bodies, bodies being nothing more than matter moving together contained in a determinate shape (*Metaphysical Foundations,* 537). And second, "simultaneous" here means not "at the same time" (for that would make circular nonsense of the intention to show that time is expressed in nature), but it means that the causal sequence is indifferent and might as well be reversed (257, 261). In other words, in this Analogy the subjective order of perception is indistinguishable from the objective one: If two substances are reciprocally each other's cause of change, it does not matter which one I fix on first as the cause: The principle tells me that whenever I perceive this indifference I may look for an interaction. But this mutual interaction taken at any *one* moment expresses temporal simultaneity.

So now we have expressions in nature of the three commonly recognized modes of time: duration, succession, simultaneity.

Much earlier Kant had respectfully criticized Aristotle for just heaping together categories, and among those that Kant mentions happen to be "at any time," "first," "at the same time" (107); these are recognizably the categories underlying duration, succession, and simultaneity. Kant has accomplished something remarkable: He has made the modes of time coherent and necessary. For it seems to me that one can think from now till doomsday and not discover why the mere concept, or even the pure intuition, of time should display just these modes in conjunction.

This question might arise: Setting aside Kant's critical purpose, did Aristotle do less? His analogue to Kantian permanence seems to me to be magnitude; to Kantian succession, development; to Kantian interaction, copresence in the cosmos. So the modes are accounted for by him as well, if not so systematically: Time is duration (and continuity) because magnitude is everywhere and always; it has succession because motion is actualization; it has simultaneity because it arises within one encompassing heavenly reference-motion.

2. *The Refutation of Idealism*

The principle for expressing simultaneity in nature was the only one to refer to space. Now "Time *and* Space" is the subject of this section; the aim is to show that for Kant time unfolds its nature only in space.

Simultaneity does indeed seem to be of all the modes the one most closely involved with space. For this mode requires an arena over which a multiplicity of mutually external beings can extend, and that is space. Kant calls simultaneity "the greatest consequence" (*maximum consectarium*) of time and writes an illuminating note on it in his dissertation *On the . . . Sensible and Intellectual World*: The union of all things that exist at one moment consists not merely in their not being successive but is an independently valid connection. Therefore time, although it has only one dimension, manages in its omnipresence to add a dimension to the magnitude of everything actual insofar as it depends, as it were, on the same point of time. For if we represent time as a line and erect the ordinates on every point of it to represent all that is simultaneous with each time point, we generate a plane that represents "the world *as phenomenon*" (para. 14, n. 5). This remarkable diagram of a spatio-temporal world integrated from successive simultaneities depicts the originary conjunction of space and time in every moment.

In the *Critique,* however, that conjunction becomes positive dependence: Time depends on space for its appearance. Kant added the "Refutation of Idealism" in the second edition (274 ff.) to show that the preceding exposition has in fact refuted the claim that existences in space are not real. The argument goes like this: I am aware of my existence as determined in time. Now all determination of time presupposes something permanent together with some change that is related to it (277), as has been proved in the Analogies of Experience. This permanence belongs to substance-in-appearance (that is, to the temporalized category) and is the same as matter (333). But matter, the result of a time-determination, can't itself be "in me," meaning in my inner sense. So it must be outside, that is, in space. Or more directly: matter and its changes cannot be in my inner sense, because they first gave *expression* to my sense of time (275). He proves that "inner experience in general is possible only through outer experience in general" (279). Of course, this argument offers comfort only through an equivocation on "outer": Kant means our *internal* outer sense, while someone seeking reassurance that the world is real will be thinking that "outer" means "independent and outside of me."

What is more, our consciousness of the existence of external objects is immediate. Should someone worry that we may have an inner sense, but only imagined intuition of space, the answer is that even such a spatial imagination presupposes an outer sense. The criteria of experience can then distinguish actual from imaginary spatial representations (277 n., 279); the criterion would be that the latter would not be determined according to the modes of time.

There we have it: Our own inner sense can be experienced only in space; time appears in space and is expressed in its realities. Space has therefore, in a manner of speaking, now become the ultimate sense.

But there is a little difficulty, parked away in a footnote (277 n.), where Kant so often buries his give-aways. The consciousness of the existence of things in space is said to be immediate. Yet we have already learned in the Transcendental Aesthetic that the inner sense is the sense of senses and that all the modifications of the soul are finally subject to it (e.g., A 99). What's more, time comprises the inner sense immediately together with the outer appearances *mediately* (50).

What does this contradiction signify? I think it signifies that the inner sense has in fact two faces, two directions of attention, as it were. One is outward into nature, where it sees its temporal nature constructed or expressed in space without any intervening consciousness. This face sees natural time directly. The other face is turned inward to the innermost part, so to speak, of the sense itself. There it receives phenomenal intimations from thought. In this inward mood (still speaking figuratively) spatial intuitions are only mediately received, for the consciousness of self—such consciousness as we are capable of—sees all appearance through its own eyes first. So the inner sense has an *immediate* and a *mediated* aspect.

Now is the moment to penetrate more deeply, so far as that is possible, into the inwardness of the inner sense.

V. Inner Sense

The best support for this two-facedness of the inner sense comes from a post-Critical essay by Kant on a prize question: "What Is the Actual Progress That Metaphysics Has Made in Germany, since the Times of Leibniz and Wolf" (1791; he didn't win). Here, in the paragraphs on space and time, he shows how we may come to know what is in our inner intuition, the intuition that underlies all our perceptions and in which the intellectual I determines the sensible I so that it may take up these perceptions and experience them, that is, may *know* them.

That I appear to myself in the sensibility is, he says, open to psychological observation. We must affect the inner sense by means of *attention*, though this may be almost painfully difficult. We will then recognize ourselves in our empirical thinking, for such thinking, as factual modifications of our representational faculty in general, belongs to the empirical representation of our condition. But then we will see that the logical I, the subject, reveals itself to our receptive sense in its spontaneity (p. 80)—and in nothing else. No further knowledge of its nature is possible.

This inward-turned consciousness shows us to ourselves as being double in yet another way than the doubling that, I argued, the inner sense displays. Kant says that "I am conscious of myself" is a thought that contains a doubling of the *I*: *I* as subject and *I* as object. How that can be is simply inexplicable, but it is an undoubted fact. (So here is yet another deep assumption of the *Critique*). The *I* that thinks and intuits is the person, the *I* that is intuited must be an "object-matter" (*Sache*), just as are sensible objects outside me. But not quite: Of this logical and now personal *I* we can know only *that* and never *what* it is. For, recall, human knowledge requires that something be given to thinking within the receptive sensibility, and thinking itself and its underlying subject, the *I*, obviously cannot be so given, for it is a functioning, *not* a sense-material. To be sure, as persons, that is, as subjects that will and act, we do manage somehow to affect ourselves directly, namely with a feeling of self-respect. But that enigmatic self-affection is the subject of the second *Critique*, the *Critique of Practical Reason* (92), since it is not a matter merely of pure cognition.

To my mind, the crucial word in the fresh explication by Kant of matters established in the *Critique* is *attention*. For I will argue in the second part of this book that if succession-time is anything, it is noting: punctuated attention (p. 206). And as so often happens to crucial insights, the one mention Kant makes of attention in the *Critique* is in a footnote (157 n.).

This note occurs in the so-called Transcendental Deduction, the one rewritten for the second edition (129 ff.). Once again I set aside the proof character of the deduction—it is meant to justify the application of the categories to appearances from a Critical, that is, knowledge-grounding point of view. I shall concentrate instead on its soul-constituting aspect, especially with respect to time. For in the B Deduction the transcendental subject and its phenomenal shadow in inner sense are set out most definitively.

1. *The "I think"*

The human subject is the "I think" that must be able to accompany all my representations (131 ff.). For if it did not, my representations would be neither thoughts, nor would they be mine. The representation of the underlying *I* is that of a spontaneous, that is, self-produced activity that is to be contrasted with the receptive sensibility (278). It is called "pure" or "original apperception." ("Apperception" is a technical word for self-consciousness.) It is conceived as a unity that works unifications of manifolds; thus, it accomplishes "syntheses."

It is a "subject" since it underlies all thinking functions and is not a mere predicate of thinking. But it is not a substance, a knowable, self-subsistent being (406–7, 413). Indeed, it is not even truly separable, since it is a mere logical form, nothing but the logical unity of the categories, completely devoid of content, the "poorest representation" (A 356, 408). It is an empty representation, of which one cannot even say that it is a concept. And it certainly has no gender. It is an I, a He, or most properly, an *It* (404). It is just all my thinking functions taken together (154).

The existence of this *I* cannot, as Descartes falsely supposed, be inferred from "I think," for not everything that thinks necessarily and really exists. The mere concept of thinking does include existence, but yields no knowledge, certainly no experiential knowledge of its subject (377). So we must say that in apperception I am conscious of myself not *as* I am in myself but only *that* I am. It is a fascinating possibility that later Kant came to give the *I* a more immediate relation to its existence, called "self-positing" (Foerster, 217 ff.). How we come to have experience of our own existence, a most crucial point, one would think, is once again set out in a footnote (422 n.); it will, of course, tell us something about the self and inner sense.

2. *Representing Time*

Before going to the central question—"What are the appearances that my *I* makes before me, in the 'inner' sense?"—it is useful to collect the representations of time in outer sense so far discussed or mentioned. For these representations are the outer side of the inner sense I have been trying to distinguish.

(a) We had, first, those appearances of time in space that determine its empirical modes: permanence, succession, simultaneity (p. 77). And what lies behind the matter that allows the modes to appear "might not be so dissimilar" from that which affects inner sense—myself (428). So I appear to myself in space. This first representation suggests a question as crucial for Kant as it ought to have been for Aristotle: Is there time in the absence of sensation? Kant seems to give contradictory answers: The real in time, which is sensation, can go to zero, and then time is empty (182, 209). But also no empty time can ever be apprehended (237). These two claims can be reconciled if there is, so to speak, an inner and outer time: Inner time would be *mere* self-appearance, in the absence of a spatial reflection.

(b) We had, next, that representation of time in pure space by means of its inscription into the pure spatial content that makes the continuity of time intuitable as analogous to the line-continuum of geometry. This inscription is a successive procedure (154), but that succession cannot be the real, that is, the natural one of (a) above.—It must therefore be *psychic* succession (155 n., again a footnote).

(c) Similarly we can represent counting, and therefore time, by a succession of points, making use of the fact that space has different point-positions (182) and can therefore image discrete magnitude.

3. *Representing the Soul*

We now come to the soul, "a name for the transcendental object (*Gegenstand*) of inner sense" (A 361)—that is Kant's way of saying "I as I appear to myself."

Kant uses almost identical expressions concerning the *I* and time: The *I* is the "standing" correlate of all my representations, just as time is the "constant" correlate of all existence of appearances (A 123, 226). This phrasing suggests an analogy: As time is to nature so is the I to inner sense. I mean that as the modes of time appear through sensation so the I appears through inner sense. I had already begun to claim that inner sense as the focus of self-appearance can be distinguished from its externalized—I mean spatialized—content (p. 79). Now is the moment to give some content to this innermost aspect of inner sense as well.

In the empirical study of the soul, that is, in psychology, self-consciousness and inner sense are identified, but Kant insists on the apparent paradox of their separation. The paradoxical consequence is that we are passive to ourselves, that we are, after all and despite previous argument to the contrary (p. 80), internally affected, and not by sensation but by our own understanding (152–53). We *are* receptive to our own spontaneity: The "I think" in its unity appears in us (140). That is how the time-sense comes to be the *inner* sense. In it the thinking subject becomes a thought-object, and, Kant says, this fact has no more and no less difficulty than how I can become my own object in general (155). In other words, how we become our own object is an aspect of the question what it means to have a representational cognitive constitution at all. The mystery of self-affection is a subset of the enigma of representationality (p. 58).

The actual contents of our inner sense are left by Kant to the speculative reconstruction of the reader. I think we must begin with the two aspects of "I think."

The logical identity that the *I* has as the subject of all thinking *appears* as the numerical identity of myself or my soul (A 363); I appear to myself as one and the same in time. From this appearance nothing follows to prove a substantial identity of myself as subject, as Kant never tires of pointing out. It would be wonderful, he says, if I could find, as if by revelation, in the poorest representation in intuition anything proving it to be a substance (408).

The inaccessibility of the actual identity of my *I* (as opposed to its inner, temporally appearing identity) is the occasion for Kant to give us an all too rare insight into the communion of subjects (A 362–63). Kant argues that if I look at myself from the point of view of another person, that is, as an object of his outer sense (of course, nothing in the *Critique* has given the grounds for *that* possibility) I must realize that I will be in his time and not in mine, for each subject's time is *in* that subject. So the other person experiences nothing of my sense of identity, the sense that always comes in the temporal form of sensibility, and he will have no reason at all to think of me as a self-identical substance. What is interesting here in a general way is that mutually acknowledged personhood must wait for the *Critique of Practical Reason* to be established (if indeed it is established there). But the particular point of interest is that different subjects have different times. For each subject has its own reason and its own world. Aristotle would have thought this madness: There is one time for us all, and one world, the cosmic community, and (at least according to the Averroists) ultimately one intellect, a view obliquely supported by the universality of time (p. 49).

So for Kant the flux of perception is backed by a standing self, "which is usually called inner sense or empirical apperception" (A 107). This inner self that *is* the inner sense is so far indeterminate, but it is now to be determined by the apperception-self (407). Recall that "to determine" means to give formal organization or to specify.

It is at this point, just where the inner sense is to be determined, that Kant introduces in two footnotes what seems to me to be the novel notion referred to above

(p. 79): the idea of an "indeterminate perception" (422 n., 429, prepared in 158 and its note). This perception *precedes* experience but is yet not a thing-in-itself. It is a sense of my own real existence, which is not thought as a category. For Kant early on made the discovery, carried over into the *Critique* (626), that existence is not a property or predicate of the existent thing but a way of "positing" it, of affirming it in thought. So such an inner positing seems to be what this perception signifies; it is "*I* think" caught in the inner act, in a sort of pure intellectual self-affection, a quasi-empirical activity in which my essence (*Wesen*, 429) is somehow given, if not thought. One might even say that this sense of existence is the cognitive counterpoint to that mysterious rational moral feeling of self-respect found in the next *Critique* (77). There is, of course, no room for such a pure perception in the critical structure; it is, in fact, a contradiction in terms. I think that Kant is driven to it by the very notion of self-appearance. It seems to be insupportable that the world should appear real to us, but we ourselves never could so appear. These auxiliary notions, marginally presented, are signals of the limits of a philosophical position, limits that it has simply by reason of saying something definite. And the final question is really always: Which enigmas can we best live with?

Thus a sense of self-existence is part of the innermost sense. But this existent *I* is also represented as thinking: "I *think*." Besides self, thinking and thinking alone is at home in inner sense (321). So the activity of thinking must also appear in inner sense as the temporal image of that originary functioning of the understanding that constitutes objects of knowledge to begin with. Such thinking is called *Nachdenken*, "after-thinking," in German (*Logic*, 1800, "Meditating," para. 120); it can be construed to mean the thinking that both *goes* after and also *comes* after the object; it is the thinking that orders and connects thoughts methodically. It is, in other words, the temporal re-thinking of the constitutive transcendental functions and judgments. Such meditation might even include the reflective thinking through which the critical terms themselves are reached by the soul in temporal thinking (316–17). Kant does not say, but it is a reasonable supposition. Such thinking might be called abstract, except that it is not drawn from, but is prior to, its objects.

Thinking, positing, and ordering are not continuous but punctuated. Attention as Kant means it, not as a general disposition of collectedness but as an act by which the soul is self-affected (156 n.), is a prime example of the "I think": self-awareness simultaneously specified into an act of thought.

One way, speculative to be sure, to think of the inner appearance of our thinking self, which is, recall, above all a unifying principle, is to think of this transcendental act as appearing in the inner sense as a series of positings of units or ones. That would be inner counting—the unity of apperception subjected to time, that is to say, appearing as reiteration. This mere attentive existence, when unfolded in the categorical judgment, is *Nachdenken*, the orderly meditation on the determinate multiplicities of the apperception, referred to just now as abstract or purely conceptual thinking. By this categorical system thinking will then grasp and unify the external manifold into those fundamental items of space called things, and these things will be the objects of empirical counting. Thus the apperception appears in self-counting internally and in thing-counting externally.

Now counting is, for obvious reasons, the prime example of attention and of the *recording* acts of attention (A 103). For counting is the result and record of the pure pulsing of inner sense and its thought-mediated projection into nature. And that brings us to memory and the phases of time.

VI. The Phases of Time

The description of time usually comes in two sorts: the modes and the phases. Kant gets the modes of duration, succession, and simultaneity by injecting temporal intuition with categorical thought. And the satisfaction of the outcome lies not only in having grounded the traditional observations, but in having located time where natural scientists look for it: in space.

The phases are much more obliquely treated; indeed to find them at all requires interpretation. This neglect makes sense if we consider that Kant is interested in this *Critique* more in establishing a human knower who sees himself in outer nature than in discovering the interior humanity of the human being. For it is that inward human being that lives through, and eventually observes, the phases of past, present, or future.

Now whether a philosopher derives the modes from the phases or the phases from the modes is, I think, a subtle but sure test of his intentions: Is the center of concern human or nonhuman Being? For Kant it is, at least in the first instance, the latter.

The phases of time are discernible in the Transcendental Deduction of the first edition. Kant probably rewrote the A Deduction because he regarded it as too psychological and too preparatory (A 98). It keeps tending to mental life as we experience it. The excuse for disregarding Kant's discarding of this version is that it shows what mental appearances Kant attended to in meditating on the pure transcendental structure of the soul. In other words, I suppose that Kant rewrote the Deduction not because the first version was wrong but because he wanted to focus on the proof-aspect of the justification for applying the categories to the appearances. Now that is just the aspect of the *Critique* I am systematically neglecting here.

The A Deduction presents three syntheses, three unifications worked in inner sense by the productive imagination. This imagination is a deep power that will later in the *Critique* be responsible for the Schematism that I took up first (p. 68). It should be said that in the B Deduction the transcendental or productive imagination is identified with the intellectual, that is, categorical, synthesis of the apperception. Kant, once again, throws this revealing comment into a footnote (162 n.; if "comical" were a term at all applicable to the *Critique,* it would be in relation to this marginalizing propensity; yet it makes sense—second realizations often come out this way). What this identity betokens is in line with the existence-sense just discussed (p. 80). It means that this underlying unifying principle of the thinking functions, the *I* that is a subject, not only itself makes a sort of appearance *in* the soul, but is also ultimately responsible for shaping the outer appearances.—The I itself is in one aspect the bridge between thought and sense. And why shouldn't that be so: Entities, says an old rule of the schools, are not to be multiplied, and neither are the nonentities, or in Kant's terms, the transcendental powers (677).

There is a temptation, which Heidegger seizes vigorously, to recognize the three phases in the three syntheses (*Kant and the Problem of Metaphysics*, para. 33). I think that project does not work out. The future, the very phase Heidegger most cares about, is not there, and the third synthesis to which it is ascribed is about something else (p. 89). I will now briefly set out the three syntheses in their bearing on time, and then say something of Kant's relation to the future.

1. *The Synthesis of Apprehension in Intuition*

This synthesis makes possible perception, which is sensation brought to consciousness. It therefore brings about the present. For Kant as for Aristotle, the intersection of time and world is the *present*, or the Now, or as Kant says, the *Augenblick*, "glance-of-the-eye." I will use "instant," since it is quite clear that Kant uses the German word as Aristotle uses "now": as the limit of a stretch of time and not a part of it (*On the . . . Sensible and Intelligible World*, para. 14; p. 43). But the problem of the present, the phase of perceptual presence, unattended to by Aristotle, is how perception, taking sensation into consciousness, can take place in the nontime of the Now. This is the problem the "apprehension" is to solve, not explicitly but in effect: how perception, the taking up in attention of a composite, can occur. (The German term *Wahrnehmung* means just that: "taking up into attention," as in *Wahren* and *Warten,* "attending" and "waiting"—nothing to do with *wahr*, true.)

Apprehension is not yet another faculty but a working mode of the transcendental, "productive imagination" (A 120). This faculty will soon be made responsible for bringing together concepts and intuition in a schema (p. 68); in the *Critique,* the Schematism of the Understanding comes after the Syntheses because they first bring out the fact that it is required. Apprehension means a "taking up" (German *Auffassen*), and so corresponds exactly to the modification of the soul it is intended to produce, perception. It is the imagination working at its "taking in" end of cognition, on the side of the intuition.

Kant says that the apprehension of sensation *fills* only one instant (209). But he also says that in counting I *generate* time in the apprehension of the intuition (182). The two propositions together indicate a reciprocity of time and sensation: Time is prior insofar as inner sense receives sensation into consciousness, while sensation in turn first determines the "outer time" (p. 76) as this and this and this instant. More simply: time receives sensation and is in turn determined by it.

The inner sense, I have argued above (p. 83), represents the *I* as a reiteration of time-units, Nows or instants. Instants therefore contain only representations that are simple unities (A 99). Kant merely posits this unity of the instant in his very brief description of the apprehending synthesis. I think it is the appearing unity of apperception that must stand behind this claim: Perception occurs in a beat of consciousness.

The intuition, on the other hand, contains all the manifold, the diversity and variety of sensations. The task of taking up this manifold into the instant is, quite naturally, assigned to the productive imagination, Kant's name for the transcendental faculty that is ultimately identical with apperception—hence it is the right faculty to work unifications in the intuition.

Kant speaks of apprehension as bringing the intuited manifold into a "picture" (A 120, *Bild*; *Einbildung* is the word for imagination). It is indeed obvious that the work of apprehension is in space; it unifies the content of space in an instant. So its materials are extended magnitudes, as required by the Axioms of Intuition (202), and neither the term "picture," nor the term "imagination," is *altogether* a metaphor. But this spatial unification has to have some temporal antecedents. A house, for example, is observed in a fairly indifferent succession of impressions before it becomes one unified representation, while in observing natural changes the attention is guided into a necessary succession (235).

In the third case, where my attention is neither randomly successive as in taking in a spatially extended form like the house, nor necessarily successive as in watching a moving body, it is objectively and necessarily reversible, because the objects stand in reciprocal relation. This nonsuccessive mode is interestingly *not* instantaneous, for while apprehension attends to one sensation-manifold it has no way of connecting it to the others to which the manifold is reciprocally related; that requires a special Principle of the Understanding, the one given in the third Analogy (257). Simultaneity is therefore not perceived but only thought—though its thought is that of a universal copresence at any instant.

The *Critique of Judgment* makes an interesting use of this fact (para. 27). Kant begins by saying what I have just argued: What was apprehended in the intuition successively cannot be comprehended in the intuition in an instant. But for aesthetic purposes the subjective imagination can work violence on the inner sense. It can cancel the time-condition expressed in the successive progression of the apprehensive synthesis by means of a "regression" that makes coexistence intuitable. The imagination is thereby enabled to receive all at once an impression of an enormous magnitude that apprehension would take a long time to synthesize. Such an all-at-once intuition is the condition for a feeling of the sublime. (See, alternatively, Makreel, 73).

In the A Deduction the prior work that goes to form an instantaneous impression is only intimated (A 120). But in the B Deduction the formation of the present is brought much more to the fore, as we will see (p. 89). Kant clearly regards the apprehension of the instant as including its antecedents. The perceptual moment trails the immediate past.

Has Kant solved the problem: "How can perception be the cutting edge of time, the present instance, when it comes about *in* time? What fills a human Now?" I think in his own terms, yes—when time is an inner sense, consciousness is constitutionally prepared for sensation. I mean that to exist as a transcendental self-consciousness that both appears to itself in time and receives sensation there, is to be, as it were, constitutionally and always in a past-supported present. Time makes perception possible. Here it is possible to articulate a difference in awareness between Aristotle and Kant, who both require an inner psychic motion to bring about time. Kant has what Aristotle lacks, a name and a structure for the fundamental psychic event: inner sense.

A final word on the perceptual present. When *all* is said and done, the *Augenblick* that is a point-Now has not been reconciled with the one that contains a perceptual unity. For it would have to be a unity that has, like a point, *no* parts, an empty unity. But then Kant, though undoubtedly concerned with giving the transcendental grounds for the present, has not really posed it in terms of that Now.

2. *The Synthesis of Reproduction in the Imagination*

This synthesis may be understood as constituting the phase of the past. Thus it deals with the transcendental counterpart of memory.

The synthesis is, significantly, named after the power that is responsible for both of the first two syntheses. In the B Deduction Kant will merge these two and call them the "figural synthesis" (*synthesis speciosa*, 151), as opposed to the intellectual synthesis stemming directly from the apperception. But that is a purely aspectual distinction, for it is in fact in this second edition that Kant identifies the transcendental imagination—now under the name of "productive imagination"—with understanding in its relation to intuition (151–52).

What is peculiarly imaginative about this synthesis? It is the transcendental counterpart of what is called imagination in the ordinary sense: the power of representing an object in the intuition even without its (perceptual) presence. Again its works are spatial; a temporal "object of the imagination" (*ens imaginarium*, 348) is not, I think, contemplated by Kant, though of course an image is, as is every representation, *in* time. I mean, once more, that there are no nonspatial images *of* time.

Except for one possibility. Kant defines the "imaginary entity," a subset of "Nothing," as an empty intuition without object; it is the mere form of intuition. Without the mirror of substance, pure space and time are thus quasi-somethings but neither is an object in the full sense (*Gegenstand*, 347); they are, in fact, nonobjects. I think that these are the pure contents of the intuition discussed above (p. 62), and so pure time as a mere "something" close to a nothing is what appears when I wipe away as much of the spatial content of intuition as possible and attend to myself as a thinking consciousness simply. Heidegger says that in the *ens imaginarium* we come on the Nothing as a something that is not among the things and is not present-at-hand (*Vorhanden*). Space and time as imaginary nothings are not intuited in the sense of being the "thematically apprehending receiving of something present-at-hand" but rather in the mode of an "originally formative giving" (para. 28). Heidegger is interpreting Kantian intuition as prefiguring ultimate primordial Time (Sherover, 150, p. 103), the originary Temporality of *Being and Time*. But he agrees that it is not so understood by Kant himself. Here is one of those fascinating cases that pose the greatest hermeneutic perplexity: whether an interpreter, in thinking beyond a text, has revealed its necessary implications, or subverted its intention. It seems to me that the latter is the case, for Heidegger's interpretation shatters the *whole* transcendental constitution of the human being, and abolishes its personhood, which stems from the *rational* will. For if time is our deepest root then the hierarchy that made us *ultimately* spontaneous, morally self-determining persons is overthrown. To save this will and with it morality, reason *has* to be prior to sensibility and so to time. The present point was, however, that the imaginative synthesis, like apprehension before it, makes configurations in space and is to be understood as dealing with spatial figures, though in time.

The word memory does not occur in the reproductive synthesis. It is treated in the *Anthropology* as a function of the "reproductive imagination" in the ordinary sense, that of a power for recalling a previous empirical intuition into the soul (para. 28). Memory proper differs from the empirical reproductive imagination only in reproducing previous representations intentionally, not randomly (para. 34). How does transcendental reproduction differ from ordinary memory?

Transcendental memory does not deal with empirical contents but is the rule that makes memory possible (A 100). The rule seems to impose a double requirement, objective and subjective. The objective rule says that appearances themselves preserve a certain steadiness in the sequence of their appearing. Human beings look like human beings, day in, day out. The subjective rule says that I do not lose once-apprehended perceptions from my consciousness. In counting I keep available the units generated (A 102) and so arrive at a number-sum. In a word, the appearances in me have an "affinity" and are therefore "associable" (A 121–22). Association is a term from Lockian and Humean psychology that Kant is here justifying.

How can this subjective rule be said to constitute the past? In the only way possible: by reproducing the objects of bygone perception as present appearances. For us, subjectively, this past is memory, which is a requirement of experience. But it is, I think, also objectively constituted, for by the affinity of all appearances, an affinity that works, so to speak, longitudinally (or temporally) in reproduction and horizontally (or spatially) in apprehension, each sensation-filled appearance is bound into its past and carries the freight of all preceding reality. In that sense, Kant has managed to ground the past as existing. It is *now* active.

Clearly the first two syntheses are "indissolubly" connected (A 102), and in their B version, the figurative synthesis, they are merged. The workings of this synthesis are most purely recognized in the successive generation of configurations in space, for—nearly—every appearance in intuition is apprehended as an extensive magnitude (204), as has been pointed out. Kant does not attend to the subjective aspect of this newly defined synthesis but one may speculate: In it the apprehension of the present and the reproduction of the past are much less distinguishable. Instead of the separable specious present of the A synthesis, which was an instantaneous impression, we have a perceptual instant at the forefront of time, an instant bound, as I said above, into *all* its past, immediate and remote.

3. *The Synthesis of Recognition in a Concept*

If the first and second syntheses constituted a present and a past, is the third not likely to give us the future, as Heidegger intimates (*Kant and the Problem of Metaphysics*, para. 33 c)?

I think the Synthesis of Recognition is not a phase at all. I think that "recognition" here means neither primarily *re*cognition, some expectation of cognizing again what was known in advance, nor "reconnaissance," a reconnoitering of the horizon within which knowledge is possible (Heidegger). It refers rather to the acknowledgment, validation, and reviewing of evidence with respect to credibility. This is the first dictionary meaning in Latin and the meaning "recognition" often has in German and English, as in "recognizing a speaker" or "recognizing credentials." It is Kant's custom to make the last category or the last principle deal with the relation of knowledge just established to the knower, and the last synthesis does just that. Kant says that the word "concept" in the title leads us to note that an act has to take place, one that unifies the intuited manifold and its imaginative reproduction into one representation (A 103). But with this conceptual unification—and really as its precondition—comes self-consciousness, whether we notice it only weakly in its effects or more strongly in the generation of the complete representation as it occurs. And without this acknowledging consciousness, this apperception, no knowledge of objects is possible. Thus the third synthesis validates the previous syntheses as being *mine*.

In recapitulating the syntheses, Kant assigns them in order to sense, imagination, and apperception (A 115). The title of the syntheses of recognition might therefore have been completed by the words "by the apperception" rather than "in a concept." The "first and highest principle" of our thinking is, as usual, set out in a footnote: "All different *empirical consciousness* must be connected in one unique self-consciousness" (A 117 n.). The last synthesis thus relates perceptions and memories to the *I*, that "thoroughgoing identity of oneself in all possible representations" (A 116). It is a solution to David Hume's problem concerning the preservation of personal identity from moment to moment (p. 177).

Yet even if the three syntheses of the A Deduction won't yield the phases of time completely, there is a way of reading the *Critique*—a deep one—that might see the whole critical enterprise as being time-informed from the very beginning. The *Critique* is, after all, devoted to the grounds of the possibility of real knowledge. Now possibility, unless it is taken in the most formal logical sense, namely the requirement that a concept should not contradict itself (624 n.) has, it seems, some reference to the future. "Real" possibility is grounded by a transcendental Principle of the Understanding, the "postulate of possibility." It says: "What agrees with the formal conditions of experience (according to intuitions and concepts) is possible" (265). Sherover, in his explication of Heidegger's interpretation, dwells on the futurity of the possible. The postulate enables the understanding to "look into the limitations of the future and see the kinds of things that may eventually be rendered present" (269). He also dwells on the priority of the possible in Kant as read by Heidegger.

It is not easy to follow this argument. The postulate (along with similar pronouncements by Kant) does not seem to me to be projecting possibilities except for the empirical use of scientists. From the philosophical point of view Kant seems to be giving grounds that are ever-present in our cognitive constitution. The futural interpretation seems to confuse the pre-temporal transcendental acts that constitute the natural world with the research projects made possible by it. In fact, the "anticipation" in the principles so named induces Kant to comment that it may be found strange that what is most adventitious, sensation, is to be so firmly expected. But, of course, the point is precisely that what is thought of as coming *a posteriori*, from beyond ourselves, is grounded *a priori*, in what is from the very first within ourselves. Thus one might in fact be justified in thinking of transcendental anticipation as *abolishing* the future and its indeterminate possibilities. Moreover, the priority of possibility over actuality seems to me hard to detect in the postulate of possibility. It doesn't say that possibilities constitute reality (Sherover, 284), but the converse—that the formal conditions of experience first open up possibilities (and impossibilities). As if to confirm this argument in the reflective dictionary of the *Critique* ("The Amphiboly of Reflective Concepts"), Kant insists that the possible and impossible are not the highest transcendental terms because they already express a division, that of the concept of an object in general (346).

4. *The Future*

If the past and present phases get meager treatment in the *Critique*, the future is all but absent—"of course," we might say, for the *Critique* is about knowledge, and there is no knowledge of the future: "a special fundamental power of our soul to *intuit* the future ahead of time (not only to infer it) . . . those are concepts whose possibility is quite groundless because it cannot be grounded in experience and its known laws" (270).

On the other hand, the *Critique* is about natural science, and therefore it is about prediction concerning nature. In fact, one might say that the Principles of the Understanding tell us what general relations we may always expect to find in nature, and that they are thus the grounding rules of scientific prediction. But again, such prediction is really the opposite of knowing the future or being concerned with it *as* futural, as a realm of possibilities: It tells us what observations will confirm the empirical laws discovered in the past and thought to hold for all times.

It is interesting, to repeat the point, that those principles called "Anticipations" (a term borrowed from Epicurus: *prolepsis*, 208) are least of all concerned with the future. They say, as do all the other principles, what I can expect to find because my cognitive constitution has so pre-ordered it. But they particularly establish the instant as the temporality of sensation (p. 85).

Anticipation in quite a different sense is introduced into the discussion of time in the *Critique* by Cohen in his *Logic of Pure Knowledge* (1902, "Judgments of Mathematics: The Judgment of Plurality," para. 11). Cohen engages in a critique of Kantian time (referred to before in the introduction to these sections) that starts from the fact that time needs to be categorically determined before it can display any of the modes of time. He then points out that these modes yield only *past* succession and *present* simultaneity. But one has to ask, he says, if it is not "the mode corresponding to the *future* that expresses the *original* tendency of time." The name for this mode is to be "sequence," and number is its proper expression. The reason is that the number-sequence is to be thought of in terms of the sequencing *activity* and of the last term that is its goal. Sequences are generated by anticipating what must necessarily follow. But "such anticipation is the proper, the fundamental act of time."

I observe three things: (a) Cohen recognizes that the future is, on the face of it, missing in Critical time, but unlike Heidegger he does not try to construe it into the text. (b) Cohen explicitly recognizes the identity of time and numbering. (c) He, like Heidegger after him, regards time as essentially futural, though unlike Heidegger, he finds the futural model in an ordered plurality, in a number-sequence; this model is appropriate for him because he had already argued that time must be a category (p. 56). Cohen's attempt to introduce the future into the Critical enterprise is, if not persuasive, at least illuminating.

If the future is understandably absent from the realm of natural necessity, one might expect to find it in the realm of personal freedom, in the *Critique of Practical Reason*. Here Kant distinguishes not only law-determined natural necessity but also past-determined human necessity found in appearance from the freedom of a human being as a thing-in-itself (95–96). For the future would seem to be the phase of freedom, the phase in which the "will" expresses the very meaning of its name. Recall that Aristotle, who precedes Kant in his lack of interest in the future, considers propositions about it indeterminate with respect to truth because of the unpredictability of deliberate human action (p. 53). But Kant says nothing at all. To be sure, there is a paragraph on the ability to see into the future in the *Anthropology* (para. 35), but it is about practical prudence only.

Nonetheless, there is one department of human knowledge in which Kant concerns himself with the future a great deal—history. One might not off-hand think of history as the focus for the future, but Kant raises the pertinent question in his characteristic way: "How is it possible to have history a priori?" For such a history alone could foretell the future ("Renewed question: Is the human race in the course of a steady progress to betterment?"—*The Contest of the Faculties*, 1798, paras. 1–2). He answers that a predictive history can occur if the prophet himself produces the events he predicts. He cites the prophets of the Hebrew Bible.

It might be said that he himself writes predictive history in this spirit. The essay "Idea for a Universal History with a Cosmopolitan Purpose" (1784) predicts the progress of the human race, through man's "*unsocial sociability,*" by means of which nature develops human capacities and makes possible the eventual establishment of just civil society and of its condition, international peace (fourth proposition). In the ninth and last proposition Kant says that such an a priori history is aimed at furthering this political purpose of nature. (Incidentally, the "unsocial sociability" is remarkably close to Madison's theory of factions in the tenth *Federalist,* 1787.)

In this manner Kant ranges from "The Conjectural Beginnings of Human History" (1786), to "The End of All Things" (1794), which deals with time and eternity, with the way in which time-determined human beings think of "the last day" of judgment and the eternity that follows: They imagine it as the ceasing of all motion and of nature become rigid and petrified.—In heaven or hell, Kant says, the hallelujahs or lamentations are ever-unchanging.

But Kant's interest in the future as a terminus of history is not concerned with the individual human capacity for futurity. And even in the *Anthropology*, which contains many acute observations on time-affections like boredom and amusement, *Lange Weile* and *Kurzweile* (para. 61), as well as on time-capacities such as prevision, divination, and prophecy, the subject is taken empirically and descriptively. It may be well and truly said that the future is not a philosophical issue for Kant.

To be sure, near the end of the first *Critique* the question is posed: What may I hope? (833), or more specifically: "If I so conduct myself as not to be unworthy of happiness, may I thereby hope to obtain it?" (837). But the answer is given in terms of what I must suppose with respect to a future life in the world beyond, not in the this-worldly future (839).

∞ ∞ ∞

The *Critique of Pure Reason* is an inexhaustible book that gives a double pleasure: One is the settling pleasure that comes when things fall well into the place assigned to them in the system. The other is the unsettling pleasure that comes from being taken to the edge of the abyss. In the study of time this latter character of the book comes out most poignantly, for time in Kant's *Critique* is neither a subject along the way, as it is, say, in Plato's *Timaeus* and in Aristotle's *Physics*, nor is it the issue itself as in Hegel's *Phenomenology* and in Heidegger's

Being and Time. Thus Kant's treatment is neither *episodic* nor *thematic*; it is instead *pervasive*, and that means that to find Kantian time we must look everywhere in the book while focusing on just one of its topics—a way of working conducive to gratifying discoveries.

What, in sum, are these? The inner sense has an inward and an outward aspect. The outward aspect is represented in the space of nature, where it is first determined as having the modes of duration, succession, and simultaneity. The inward aspect gives appearance to the thinking *I* in generative acts of attention that are a kind of original numbering motion, as well as in discursively ordered abstract thinking. The phases of time arise in tandem with the stages of the synthesizing that bring about real knowledge, which is concepts applied to appearances. The future is therefore missing, since there is no synthesis of mere possibilities.

In Aristotle's treatment of time the unanswered question was: How is the psychic motion of counting, the counting that produces time as the number of motion, to be explained? That is just the question to be put to Kant: What is the punctuated fluidity of the inner sense? Is it not a motion that is itself the hidden condition of time? And it is no more answered by him. Or by anyone.

One last comparison: Neither for Aristotle nor for Kant is time, though its seat be deep in the soul, deeply connected to the passions or feelings. Both philosophers are well acquainted with the noncognitive affections of the soul, but neither think of time as being at the root of human passion. Aristotle consequently puts time in his treatise on motion, the *Physics*, and Kant puts it in the *Critique of Pure Reason*, where he means sharply to distinguish knowledge from feeling (66, 829 n.). We may conjecture that there must be a view of time that absorbs it into the fabric of the *whole* soul. And that view will now be taken up.

Chapter Three

Plotinus and Heidegger: The Grounding of Time

The "heads up" signal that demands to be heeded for the parallelism I am about to present is the startling sameness of key terms and emphases in the two authors. Let me name some of them: temporalizing, ecstasis, dispersion, fallenness, future. They will be the focus of attention in this chapter.

This identity of terms is not, I think, a mere, meaningless accident. It is rather the case that for all the abyss of intention and for the more than a millennium and a half of time that separate Martin Heidegger from Plotinus, there is at least one point of commonality, and that a deep one: Both philosophers think that for time to become intelligible it must have a *ground*. Some source, higher or beyond, more divine or at least transcendent, must account for human temporality, which consequently is fatefully ingrained in, or better *is*, human Being or Existence, in all its defectiveness. It is that common starting point that yields a common vocabulary.

Nor can I think that Heidegger silently cribbed from Plotinus. That would anyhow be mere creative plagiarism, from which nothing much follows. But even in the case of Hegel, from whom he takes more extensively than he acknowledges, there are plenty of credits. It seems, however, that Heidegger did not pay much attention to Plotinus, who appears but once, in passing, in those of his books that I am using and only marginally in others.

These underlying texts are, I should say now, *Being and Time* (1927; translated by John Macquarrie and Edward Robinson, 1962) and its completing sequel, *The Basic Problems of Phenomenology* (lecture course, 1927; book, 1975, translated by Albert Hofstadter, 1982); for both works I cite the standard German pagination. For Plotinus I used mainly the annotated text and translation into German of *Ennead* III 7, "On Eternity and Time" (c. A.D. 270) by Werner Beierwaltes (1967). This edition is worth its weight in gold, and I'm much indebted to it. *Enneads* I 5, "Whether Happiness Takes Increase in Time," and IV 6, "On Sensation and

Memory," are also necessary reading. (The posthumous publication of Plotinus's meditations was called *Enneads* because his pupil Porphyry collected them into six sets of nine; "ennead" means a "niner.")

Nothing needs to be said here of Heidegger's life, but my favorite fact about Plotinus is well worth pondering. He made a laborious but unsuccessful effort to reach India, the source of mystic wisdom; to my mind this failure was fatefully lucky, since the continued effective separation of the Greek and the Indian traditions makes for crisper distinctions than we would have had under an early confluence—or so I imagine. In any case, the concentration on the Western tradition maintained in this book would probably have been impossible had Plotinus reached his goal and returned safely. The "history of time" would have been—unimaginably—other than it is.

A. PLOTINUS

I will now go on to the delicate business of retailing in brief what Plotinus thinks about time. The chapter in question, *Ennead* III 7, was rightly entitled by his editor "On Eternity and Time," even though in common parlance "time and eternity" roll more smoothly off the tongue, even in Greek. So too, in philosophical considerations, eternity, that is, true eternity as opposed to "sempiternity," which is current time in its endlessness (Boethius, *On the Trinity*, 4, 72), is usually approached by a *via negativa* or by analogy with temporal features, but above all, in terms of the negative analogy to the passing, pointlike Now of time. This eternal Now is both unmoving and full, it is the *nunc stans*, the standing Now, that timelessly embraces all time. Though Plotinus too will use such language (3, 28; such numbers will from now on refer to the subsections and lines of *Ennead* III 7), all his effort is to derive time from eternity and as far as is humanly possible, *to explain time through eternity*. That is what I mean by saying that he *grounds* time: Time is not an original or originating being or mode but a fall-out from something higher. I might observe here that anyone sensitive to meanings will have cause to object that a higher origin is very infelicitously called a "ground," which should underlie rather than overhang its effects. My excuse is that in philosophical speech talk of dependency and derivation is often intermingled with terms of grounding and underlying. What matters here is the priority of the source, though I will also need to draw attention to the metaphorical direction of these relations.

Plotinus likes to use the analogy of model (*paradeigma*) and likeness (*eikon*) with respect to the relation of eternity and time. He is following the famous passage on time in Plato's *Timaeus* (37 d 6 ff.; p. 4), which he believes—in a less earnest author I would be tempted to say, pretends to believe—adumbrates and prefigures his theory. For what characterizes him as what we call a Neoplatonist is in this case just this: that he fills in metaphysical derivations where Plato's Timaeus (not necessarily Plato himself) speaks in figures. I am suggesting that by converting Timaeus's myth into an ontological framework, he makes time undergo a conversion too, from the merest name for something else to the deepest condition of human beings.

Plotinus, then, does his best to keep from using privative time-words. A nice indication of this effort is to be seen in the use he makes of the etymology of the Greek word for eternity, *aion*, that occurs in that passage (p. 6). The etymology for *aion* that Timaeus has in mind is "un-going" or "not-going," that is, unmoving. Plotinus, however, will quote and seize on Aristotle's construal (4, 42), who makes *aion* mean *aei-on*," being-forever," or perhaps "being unconditionally" (*a-ei* could mean "not if," but this is not explicit: *On the Heavens* 279 a 27), though Aristotle is talking about infinite or sempiternal time.

Plotinus does not, however, quite succeed in eradicating privative language from the description of eternity: It is, besides being abiding and all comprehensive, also a timelessly present presence in which everything already is; it has neither "was" nor "will be." But Plotinus can specify the negative description so that it begins to intimate the derivation of time from an inherently more intelligible eternity. What eternity most specifically lacks is futurity: It has no "appetite for things to come" (4, 34); it does not fall from the seat of—present—Being (4, 21).

And that intelligibility is the result of the derivative nature of eternity itself. To speak of Plotinus is to speak of *hypostases*, a multiplex word. One might be tempted to translate it fairly literally, as "substance," and that would be right in a crucial respect: a substance is what stands on its own. The Plotinian hypostases do have a sort of independence, though they eventuate from a higher being that contains them potentially. They themselves are the actualizations (*energeiai*) that flow from the prior being. Note a consequence of interest in comparing Plotinus with Heidegger: for Plotinus potentiality or potency (*dynamis*) is *in a certain way* above actuality (*energeia*); for Heidegger that is unambiguously the case.

The hypostatic relation is asymmetrical. Each hypostasis is forever somehow enfolded in its superior, but that superior has no relation to its alienated and released offspring. Hence the direction of all striving to return to what might be called authenticity, the condition of genuine Being, is upward. The Plotinian motion of salvation is unidirectional, to the hypostasis above (VI 8, 16, 10).

To become more specific: Plotinus himself speaks of three great hypostases: The One (*hen*), the Intellect (*nous*), and the Soul (*psyche*). The One is clearly not a true hypostasis since, as ultimate font and origin, it is itself ungrounded. (Plotinus himself uses the Greek word for grounding, *hidryein*, IV 7, 9, 14, figuratively false though it is.) The One comes into this chapter only as a final source. Its hypostasis—and here the word might be translated as "sub-station"—is the Intellect. The Intellect comes out of the One insofar as the One has a certain self-regard, engages in self-reflection, and so undergoes a kind of diremption. That atemporal motion of separation foreshadows the origination of the third hypostasis, the Soul.

Let me now state the crux of Plotinus's treatise on Eternity and Time. It is this: *The Intellect is identical with Eternity. The Soul is identical with Time. As Soul falls out of the Intellect, so Time falls out of Eternity*—only what I have stated as an analogy is really an identical event.

What, to begin with, is the common character of Intellect and Eternity? How is their identity to be understood? Intellect, still close to the One, is all together, collected not in the supervening sense but originally. It possesses stillness or stationariness (*stasis,* 2, 20). One might say of the Intellect that it is *in*tuition, and in a triple sense. It betokens collected intensity, as in "*in*flammable"; it refers to the beholding of something, as in the word *in*tuition itself, which means at-sight (cf. German *An-schauung*, p. 61); and it connotes undistracted looking-within, as in *in*sight. Intellect is the thought that thinks or views itself, the thought in which thinking and thought are immediately together (3, 4). But most characteristically, it is instantaneous, unextended, undistended; it is that which does not distance itself from itself (V 8, 9, 19). And all that has been said of Intellect applies to Eternity: It is still and unchanging; it is Stationariness Itself, it "abides in the One," that is, it stays within its unity; and it is comprehensive. But above all it is dimensionless, unextended, distanceless—"undistanced" (*adiastaton*) is Plotinus's term (2, 32–36). One might say that Intellect and Eternity are alike statically instantaneous, standing Nows. But they are not just alike; they *are* each other. In Plotinus's words:

> The life that in its being is involved with Being, a life that is both all at once and replete and undistanced in every way—that is what we seek: Eternity. (3, 36)

This life is the life of the Intellect, and life is what the Intellect essentially *is*. So Eternity *is* Intellect. It is an astounding proposition, one here intended to prepare the genealogy of time.

The next question should then be: How does Soul come out of Intellect? Intellect eventuated from the One by a kind of shining-out that was also a "standing-off," an *apostasis*, a defection, to use Plotinus's own word (VI 6, 1, 1). I might remark here that these *sta* words, *stasis*, *adiastaton*, *apostasis*, and soon *ekstasis*, are central to the Plotinian vocabulary (p. 100). For remotion, distancing, from an originary *stasis* or stand-still is *the* activating motion in his ontology.

The Soul "comes down" from the Intellect in the same kind of intellectual motion that brought Intellect out of the One. It arises for our understanding as an actualization of a potency in the Intellect, and for our figurative imagination as a copy now twice removed from the One. So there is a second fall (V 1, 3, 7). And again it has to do with dispersion. As Intellect fell from the One and became "one-many," one that *was* many, Soul falls from Intellect and becomes "one-*and*-many," one that is *also* many (IV 1, 2, 53). The Soul is the Intellect as it has longed for self-alienation and is seduced into the sensible world (IV 6, 3, 10). In taking on alien matter, the Intellect, now as Soul, is no longer one with its own thought and does not think immediately. Rather it now thinks through a distance. The Platonic term Plotinus finds handy is *dianoia*, which means literally "thinking through," as opposed to *noesis*, distanceless, immediate thought. In accordance with the priority of potency, *dianoia* is called the *energeia* of all *noesis* (I 1, 9, 22). In school language, the soul's thinking has an *object*, a presence *before* it. Heidegger's misconstrual of antiquity notwithstanding, the highest intellection for the Greeks is a presence that has no *prae*, no positional "before" in it, an activity perfectly melded of "subject" and "object"—*noesis noeseos,* "thought-activity of thought-activity" (Aristotle, *Metaphysics* 1072 b); the object of or for thought enters the picture only at a second remove from the source.

The Soul, pendant betwixt and between the intelligible and the sensible realm, had made for itself "a place of dissimilitude," in which to exercise "a certain unquiet potency that wanted forever to translate the things seen There into something else and did not wish the collected All to be present in itself" (11, 23). Thus the Soul is Intellect that has allowed itself "to be wrenched away" from its silent unity to show its "busybody nature" (11, 15). "Busybody" is the word surely suggested to Plotinus by Plato's use in the *Republic* where it is a harsh indictment: The unjust man is one who fails "to do what it is his to do and not be a busybody" (433 a).

This desire for self-exceeding business, for noise and novelty, now provides the answer to the last question: How did Time "fall out of" (11, 7) Eternity? The answer is, of course, that as the Intellect *is* Eternity so the Soul *is* Time. Here is how this identity is to be understood, how Eternity comes to be the precursor of Time, and Time the life of the Soul. The Soul imitated the intelligible world and made a likeness of it, a sensible world that moved not with intelligible but sensible motion. But, and this is the crux, *first* she did this to *herself*—and here is the intelligible moment of her fall: She "temporalized herself" (*heauten echronose*) "making instead of Eternity, Time" (11, 30). And so the world moves in time, since the world moves in the Soul and hence in the Time that is the life of the Soul. For there is no other place for this "All" to be than in the Soul. I might note that here place is consequent on time, for time makes place possible: By its self-distancing it makes a place for place (IV 3, 9, 22; for the order of priority in Hegel, p. 24; in Kant, p. 62; in Heidegger, p. 108).

Perhaps this is the moment to mention that although Plotinus has throughout been speaking of a World Soul, human souls are entirely analogous in their temporal condition. "Is Time in us?" (13, 66). Yes, but not as a psychological factor, or even a transcendental element, as for Kant, but as our *Being*.

Thus the soul temporalizes itself, or better, the Intellect temporalized *is* the Soul, because the thinking that deals with otherness and multiplicity is drawn out and is *diastatic*, "distantial." Time is the *diastasis*, the self-distancing of the soul, that non-spatial extendedness which shows itself as our *dianoetic*, discursive, *sequential* thinking. For succession belongs to time as well as does extension, a before-and-after that is forever passing, and that tends toward infinity, never achieving wholeness (11, 54). Time *is* discursive thinking.

This temporalized soul is essentially *futural*. The being and nature of Time shows itself in the future. For it was the restless, unquiet longing for something else and something more that made the Soul cast itself into Time, where Time can now be defined as "the life (*zoe*) of the Soul in its transiting motion from life-phase (*bios*) to life-phase" (11, 43). And that transition moves into the future, taking in ever new time and turning it into bygone time. For the nature that made the Soul temporalize itself was the desire for ever additional acquisition (11, 53), and this desire expresses itself in a perpetual search—a future-directed effort.

Finally, the reversal of the fall into Time, the reunion of Intellect and Soul (11, 1), its recollection (*anamnesis*) into unity, its turn-about (*epistrophe*) and return, and its road upward and back to the origin (*anagoge*)—in sum, the detemporalization of human thinking—gives rise to a couple of terms I want to note here briefly, for we will hear versions of them again from Heidegger.

Ordinary memory (*mneme*) is a mere Intellect-analogue; through it the diastatic Soul holds fast temporally to what it no longer has (III 6, 2, 44), while as Intellect it had everything at once. In recollection, an atemporal event, it recovers the thought it has always possessed (IV 3, 25, 33). The state of true happiness (*to eudaimonein*) similarly is not in time nor does it increase with time; it is an unchanging state (I 5, 1). It is love, *eros*, that has the power to collect the Soul, not allowing it to remain dispersed. From the "smudged gleam" of the sensible world, *eros* rises into a "great light" (VI 7, 33, 30), and the Soul withdraws itself from that world as *eros* works a "turn-around." While the temporal thinking of the soul does go on noiselessly (*apsopheti*, 12, 3) the soul's real return is to its—atemporal—former eternal silence (*hesychia*). Plotinus abhors noise, multiplicity, extension. Self-collection is, as I have noted, not a forward but an upward motion. The Soul's way up leads not out of but to herself (VI 9, 11, 38). The return happens "suddenly" (*exaiphnes*, e.g., V 3, 17, 28; see also the all-important *Seventh Letter* of Plato, 341 c–d). It is a moment of insight in which seeing and what is seen are again merged, a moment that, as in the blink of an eye, unifies Being and thought in a "different way of seeing"; Plotinus's term for this state of sudden, silent reunion, in which the human soul is rapt away (*harpastheis*) and enraptured, is "ecstasis" (*ekstasis*, VI 9, 11, 6; this reference for all these terms is, significantly, to the very last treatise of the *Enneads*). *Ekstasis* is the ultimate *stasis* word. It means getting "out of" *diastasis*, out of self-distancing, and turning back into the *stasis*, the stationary stillness of the One. For Plotinus, *the aim of ecstasis is stasis.*

B. HEIDEGGER

I must now do something that is as difficult as it is dubious: give an abbreviated and highly selective reading of Heidegger's philosophy of time, interleaved with the Plotinian analogues and antitheses. The dubiousness of this enterprise lies mostly in the fact that time is what *Being and Time* is ultimately about, so that seizing on some elements relevant to my thesis may skew the account; and the difficulty lies in the temptation to slight the radically different context within which Heidegger's theory of time arises. So I will attempt to be as balanced as brevity permits.

Let me first repeat the thinking behind my thesis: It is that Plotinus and Heidegger alone among the well-known philosophers I've read, although poles apart in their beginning, motives, and ends, have in common the particular views concerning time laid out here. Not only do we, in focusing on these commonalities, get a tighter grip on the differences of the contexts from which they are abstracted, but here is what is more to the point: We learn from the similarities and differences what is the order of profundity with respect to the questions to be asked in the inquiry concerning time. Moreover, in marshaling the cognate answers we see what aspects of time seem to survive the most diverse approaches and present themselves no matter what agenda might drive the inquiry. Finally, here as everywhere, my thought is that to think about time is to think about one's life, and to come to some clarity about time is to live somewhat differently (p. 161).

Let me also now repeat my thesis itself: It is that Plotinus and Heidegger, alike and alone among the great philosophers of time, think (1) that time *is* human Being, (2) that time needs a ground or an ontology, and (3) that time so construed evinces specific features such as a "diastatic" or *ekstatic* nature, a privileged future phase, and an element of fallenness.

And now to Heidegger compared to Plotinus. In accordance with its title, time, or rather Temporality, turns up in the second part of *Being and Time*. "Temporal," as used in ordinary parlance, means "secular" as opposed to sacred. Heidegger uses the Latin term, Temporality (usually rendered with a capital T in English), to mean "primordial time" or the ground of possibility of time, that "upon which" time is "projected"—that which makes it possible for us to understand it. Temporality is therefore an ontological term that raises the question how time can be constitutive of human Being. Plotinus will, as we know, once, but then crucially, use a Greek verb he has himself made up: *chronoun*, whose translation is "to temporalize." (Aristotle, to be sure, often uses the verb *chronizein*, whose perfectly ordinary meaning is "to spend time," but whose accurate translation would also be "to temporalize.") Recall that for Plotinus primordial Time is Eternity.

As the term for human Being Heidegger uses *Dasein*, literally, Being-There, or better "There-Being." In ordinary German it means everyday existence, as in "the struggle for existence." In Hegel's *Logic* (89–90), *Dasein* is the word for the dialectical result that follows on the collapse of labile Becoming into a unity: determinate being. For Heidegger *Dasein* names that unique kind of Being, our own, that has existence rather than mere thinglike extantness. (For a note on Heidegger's specific uses of *Dasein,* see *Basic Problems of Phenomenology*, appendix). *Dasein* "ex-ists," literally, "stands out." It is thus the being that discloses space, or more deeply, that discloses a world about itself; it is the cause of that opening up or "unlocking" (*Erschliessen*) that makes things shine out. Now strangely enough, for Plotinus too "There" *(ekei)* is a word of import. Unlike *Da*, however, which means *both* There and Here, *ekei* means only There. The second treatise of the fourth Ennead on the Being of the Soul begins with the words: "In the intelligible cosmos is the true Being. Intellect is the best part of it. But souls are There too. And because they are Thence they are Here" (*Ennead* IV 2, 1). "There" means the atemporal realm, the realm of true Being—just the opposite of *Da*. I might add that for Heidegger transcendence is achieved on the ground of Temporality, not in the sphere of the atemporal Beyond (366); the explanation is in terms of the temporal horizon (p. 106).

Dasein alone admits of having a meaning, that is to say, of being a ground. For ground becomes accessible only as meaning (152; such numbers from here on refer to the German pagination of *Being and Time* always cited in English editions). "Meaning," speaking abbreviatedly, "unlocks" the ground of the possibility of Being (324).

The first part of *Being and Time* sets out an analysis of the features that characterize *Dasein*, that one being that has understanding, that makes an issue of unlocking its own meaning (12); the book therefore presents itself as an example of what it means to be human. The same features so disclosed, are then, in the second part, grounded in primordial Time or Temporality; they are levered, as it were, into a more basic ontology. I shall take up those features, called "existentials" by Heidegger—the chief ones, as it happens—that are relevant to my thesis, and I shall present them as they appear in their Temporal setting.

The fundamental, primordial feature of *Dasein* is Care (*Sorge*). Human beings exist as caring in the basic mode of Care, which is specified by means of two prepositional prefixes in German: *Besorgen*, or "concern," which means "taking care of," as of business, and *Fürsorge*, or "solicitude," which means "caring for," as for an invalid (193).

Plotinus too speaks of Care. In fact, the Greek word *epimeleia*, which occurs exactly where one would expect to find it, where indeed I did find it (*Ennead* IV 8, 2, 27 ff.), has a nice feature *Sorge* lacks; it is—though doubtfully—linked with the verb *mellein,* which indicates futurity in Greek, as in "I'm expecting to. . . ." *Epimeleia* turns up in the treatise on the descent of the soul into the body, where Plotinus asks how the soul can take care of its world in the face of the condition that the higher being must never become involved with the lower. Plotinus needs to distinguish a lower, practically intervening and infecting, care from a higher "nonpractical" supervision such as allows the soul to help its world to flourish without being itself dragged down. And Plotinus believes the world to be worthy of such care: All due contempt for its bodily being notwithstanding, the world's splendor demands the reverently concerned question "what [wonders], whence?" (II 9, 16, 55). So true Care for Plotinus, although a mode of the soul's Being, is just the opposite of what it is for Heidegger. For Plotinus it belongs to the Being of an uninvolved source *beyond* the world, for Heidegger it is involvement *in* the world—"Being-in-the-world." I cannot resist pointing out that the more remote philosophy of Care may well be productive of the greater intramundial care: We know from Porphyry, Plotinus's assistant and biographer, that he was humanly attentive, kind, and generous (*Life*, 9, 17; 13, 8).

Care, in *Being and Time*, is a unity of three existentials illuminating the various meanings of Care. They are (1) the deepest existentiality as revealed in understanding, (2) facticity as revealed in "the state in which one finds oneself" (*Befindlichkeit*), that is, mood in the most revealing sense, and (3) fallenness, the state of inauthenticity par excellence. This is so far mere vocabulary, which must by now be briefly explicated in temporal terms.

Understanding, the very meaning of existence, is *Dasein*'s way of being, even before it appears as its cognitive mode. It can appear as "authentic" (*eigentlich* means proper, intrinsic, in ordinary German). Or it can be inauthentic. The possibility of existing inauthentically arises for *Dasein* because it is that one being whose Being is at issue for itself. It is, moreover, that being which can choose among its possibilities: As a being whose essence *is* its existence, and whose existence *is* its possibilities, it can lose itself, alienate itself (41 ff.). Recall that the Soul's self-alienation is the cause of time for Plotinus and that the issue of losing herself in the sensible world or returning to her true There is a governing problem for the Soul. Understanding appears in the three phases of time, once authentically and once inauthentically (Gelven, 191).

So now the moment has come to render Heidegger's grounding of time. This crux of the whole book is presented on one single page, 329. I might say here that it is a familiar way in Western philosophy: What matters most is presented most briefly (p. 67).

Temporality, primordial time, is "manifest as the *ekstatikon* pure and simple. Temporality is the primordial outside-of-itself in and for itself." The phases of time—future, present, and past—are therefore called "the *ecstases* of Temporality." Recall that Greek *ekstasis* means a "standing out." Of Temporality itself it is hard to speak, because it is the ground itself (328). However, it is said to be not an entity but "a process of temporalizing in the unity of *ecstases*." Here "temporalizing" translates *Zeitigung*, formed from the verb *zeitigen*, which in turn comes from *Zeit*, time. The verb ordinarily means "to bring something to maturity." Heidegger reemploys it in the sense of bringing time into being, of casting something into time. Recall that the Greek counterpart, made up by Plotinus and used but once (p. 99), is *chronoun*. Such verbs, including "to temporalize," are called "denominative" because they are made from a noun (or an adjective) and betoken causing something to be in the mode that the noun or adjective indicates. Probably no single point of similarity is more striking than this one-time usage by Plotinus of a word that describes the self-temporalizing of a soul alienating itself both from itself and from the unextended intellectual "motion" of its true There (*ekei*, III 7, 11, 28) to come apart into the phases of time; thus one might well say in Heideggerian language that the Soul's primordial Temporality is Eternity.

The ecstatic constitution of human Being, interpreted by Heidegger as "existence," was already foreshadowed in that very word whose literal meaning is the same as "ecstasis"—"standing out" (see 329). And the character of this constitution was read off the three existentials that comprise Care: existentiality or understanding, facticity or state-in-which-one-finds-oneself, and fallenness or deterioration. More of this in a moment.

Thus Plotinus and Heidegger both view human existence as "diastatic" or "ecstatic" (*sta*-words are as important for Heidegger as for Plotinus; p. 98). For both of them it is this self-distancing or "being-out-of-itself" that accounts for the stretching out of worldly time. In fact, Heidegger uses "stretched-outness" (*Erstrecktheit,* 410) as a technical term for what might be called the primordial span of a whole existence between birth and death; this stretching allows for the *Dasein* not to *have* but to *be* its past and so to become historical (375). "Stretching-apart," *distentio*, is a term used by Augustine to explain the simultaneous presence in the mind of the phases of time, of past, present, future (p. 122). But for Augustine "distention" is the time-producing property of which each human mind is the substrate, not the primordial structure of soul or existence. Both Plotinus and Heidegger think that for some being to be Soul or *Dasein* means being distanced from itself, being out of *itself*—self-projection (*Basic Problems,* 318).

This crucial grounding Temporality had to be set out before "understanding" could be further interpreted. Understanding is, to begin with, that existential feature of *Dasein* that reveals to it its "power-to-be" (*Seinkönnen*, 144), the possibilities that make it what it is. Recall that for Plotinus too, the potency (*dynamis*) of the releasing ground precedes the actuality of the hypostasis (p. 97). The motives are of course poles apart: for Plotinus it is a matter of preserving a generating hierarchy in which the power of releasing an actuality (*energeia*) as a hypostasis is always in the being immediately above. Heidegger wants first of all to avoid understanding human Being, "existence" in his term, as an "actuality." For actuality means enactedness, and "effectedness" (*Gewirktheit*), that is to say, it designates that by which a thing is posited as *extant* (*Basic Problems*, 123). But the question concerning human Being should not be *what* but *who* it is (ibid., 169); it should not concern extantness, mere presentness, but existence with its choices. Yet the dignity of this being, who has its Being in its possibilities, is not to be understood in terms of merely logical possibility. Rather we must think of it in terms of temporalizable capabilities or powers, as Heidegger's German term *Möglichkeit*, connected with "might," indicates—"real" possibility. Thus he says in a most revealing paragraph that Temporality as origin is richer and more pregnant than anything that arises from it:

> This makes manifest a peculiar circumstance, which is relevant throughout the whole dimension of philosophy, namely that within the ontological sphere the possible is higher than everything actual. All origination and all genesis in the field of the ontological is not growth and unfolding but degeneration, since everything arising arises, that is, in a certain way runs *away*, removes itself from the superior force of the source. (ibid., 438)

The passage turns out to be, after all, close to the *Enneads'* pervasive teaching. Interestingly enough, Plotinus uses "running" (*dramon*) and *"running beyond"* for exactly the opposite motion, the return to the source (III 7, 5, 9).

Now I want to return to understanding and its revelation of *Dasein*'s possibilities. (Of the six temporalized modes of the understanding I will mention briefly or omit those that are not to my point; they are not against it either.)

The primary meaning of existentiality is the future (*Being and Time,* 327, 337). Existentiality is revealed through understanding. So for *Dasein*, as the being who understands and for whom therefore its own possibilities are an issue, the future is primary. One might indeed have inferred that from the emphasis on possibility, which is, except in logic, future-directed. The authentic understanding "anticipates": Authentically existing *Dasein* lets itself come toward itself as its own power-to-be (336 ff.). This interpretation of temporalized understanding is suggested by the German word for future, *Zukunft*, literally, "coming-towards." Inauthentic understanding does not anticipate—it does not go toward the future, but lets the future come at it and merely, passively, expects it (p. 174).

Recall that for Plotinus too the future is the primary phase, and for him too the Soul's constitutive understanding, the *logos* (and its discursive mode, *dianoia*), is primordially diastatic, so that "by it" Plotinus will generate time (III 7, 11, 4). For him too there is a return to herself of the Soul in *noesis* and an alienation in *dianoia.* This discursive thinking is enslaved to busyness and to novel succession and is ever expectant of an ever-yet-to-be whole (ibid., 56). To be Soul means to generate "actualized discursive thinking" and with it "what is not before," that is to say, what is constitutionally futural.

Back to Heidegger: What is it that authentic understanding primarily lets come toward itself? It is death, that possibility existence has of coming to its end, an end that is most one's own to face or to evade. The second part of the book begins with the central existential feature of *Dasein*, which Heidegger calls "being-towards-death" and grounds in Care (259). Here is an absolute antithesis to Plotinus, for whom death is no great issue, certainly nothing to dread or fear—no evil, since it means reunion with the whole Soul, the World Soul (I 7, 3, 9). In any case, for him removal from life does not need to wait for death. Plotinus reserves the term *ekstasis* for that getting out of our temporal souls that is achieved "suddenly" by a well-prepared Soul (VI 7, 34, 13) when it is lifted up by the Intellect and sees not something else that is visible, but the light itself, the light that is itself the vision (ibid., 36, 19). And this ecstasy is our most proper possibility. But why, Plotinus asks, can we not stay There? Because we have not yet "come out" completely, because our bodies still hold us (VI 9, 10, 1).

But then again the two philosophers come close: While for Heidegger authentic human time is finite, inauthentic time is infinite; this temporal infinity, this endless succession, is exactly what Plotinus opposes to the simultaneously whole infinity of Eternity (III 7, 11, 54).

There is for Heidegger also an inauthentic temporalized understanding pertaining to the present. The inauthenticity of this ecstatic mode is more vividly described in *The Basic Problems of Phenomenology* (434 ff.) than in *Being and Time*. In both books a "horizontal schema," a direction-giving and delimiting horizon, is defined (365, *Basic Problems,* 436): Ecstases are not simply raptures—Plotinus too speaks of being "rapt away" (VI 9, 11, 13) in sudden ecstasis—"in which one gets carried to (*zu*) . . ." an indeterminate destination. Rather there belongs to each ecstasis a "whither" to which one is rapt away (365).

I am going to slight these horizontal schemata although in *Basic Problems* (436–37) it becomes clear that temporality understood in terms of the unity of the horizontal schemata "constitutes the general concept of Temporality" and is therefore our avenue of approach to primordial Time. The schemata are the ultimate ground on which understanding casts its projections, and its horizons are the boundaries ("horizon" is derived from the Greek verb for setting bounds). From these, in our finitude, we take our direction: the uses *Dasein* expects to make of equipment offered by the world (future), the purposes it has in a given situation (present), and the mood in which it faces what it finds (past).

I will then set aside the other schemata and say a little more only of the horizon with respect to which the present is defined. It is called *praesens*, "being before," and it makes possible the commerce that existence can have with extantness, in other words, our life alongside of *things*. It is the locus of perception, in non-Heideggerian terms (*Basic Problems,* 449). In this inauthentic ecstasis, understanding engages in "making-present," *Gegenwärtigen* (328). The word is formed on *Gegenwart*, "the present," which Heidegger etymologizes as "waiting alongside." This mode of understanding is thus a falling inauthentically among things that are not existences. *Praesens* is also the horizon for the temporal mode that Heidegger despises as now-time (423 ff.), and Plotinus denigrates as succession (III 7, 11, 37).

A word on despising: the hierarchical Plotinus is perfectly candid, in a kindly way, about superior and inferior modes of being. Heidegger, on the other hand, often uses terms that are in ordinary speech denigrating, but abjures any such intention (169): "Inauthenticity" itself, as well as such terms as "fallenness" (*Verfallen*), and "chatter" (*Gerede*) are examples. I might mention here that both philosophers have a special respect for silence. For Heidegger *Dasein* is authentically itself in "silence-keeping" resoluteness: "Authentic being-one's-self *as keeping silent* does precisely not keep saying 'I–I'" (323). For Plotinus the soul rested in quiet before she temporalized herself, and even when temporal, her motions are still noiseless (p. 100).

The authentic present is called the *Augenblick*, "the glance of the eye." It is connected with resoluteness (*Entschlossenheit,* literally, "un-lockedness," 328). I cannot resist pointing out the martial connotations: *Coup d'oeil* and *resolution* are named by Karl von Clausewitz as the two virtues indispensable to the mind under arms (*On War*, III). "Glance-of-the-eye" is not to be confused with the inauthentic Now. It means rather that *Dasein* is ecstatically "rapt away" to a confrontation of the situation and its possibilities with full existential understanding of itself (338). Plotinus's sudden rapture to the There, though described as a moment of vision, is, on the contrary, *out of* rather than *into* the world. For him there is a There there.

With respect to the ecstatic past, understanding engages authentically in "bringing back" or "repeating" (*Wiederholen*) and inauthentically in having forgotten (339). Heidegger's "bringing-back" has certain formal features of Plotinus's *anamnesis*, "recollection" (III 7, 1, 24; IV 3, 25, 33), by which one may atemporally regain what was once one's own. "Bringing back" means going back into the possibilities of a *Dasein* that has been there, *da* (385); formally that holds for Plotinian recollection as well, but there the similarity ends.

And finally we come to "fallenness" (*Verfallen*, with a sense of ruination, deterioration), which, together with "state-in-which-one-finds-oneself" or mood (here omitted), completes the structure of Care: Its Temporality is altogether and only inauthentic, and belongs primarily to the present, not to the present of the "glance-of-the-eye," which resolutely encompasses possibilities, but to the present of mere curiosity. Curiosity does not dwell with extant things to understand them; it is aware of no possibility but is greedy for what is most inauthentically futural, namely novelty, which it desires as an actuality (347; curiosity is *Neugier*, literally "new[s]-greed"). Actuality, as the end of possibility, is, of course, a low, derivative mode of being. This actual presence, which no longer allows for a possibility of Care and a proper future, is the existential-temporal condition for the possibility of distraction (*Zerstreuung,* literally "dispersion"). It should be emphasized that for all these temporalizations, "Temporality is essentially falling" (369).

For Plotinus too there is falling, sometimes "falling out": Time falls out of and down from Eternity (III 7, 11, 7). Souls that have immersed themselves in matter are impure and fallen (I 8, 1, 19). Fallenness is the other perspective on the generous unfolding by which the hypostases come to be—as they are released, so they are separated and cast out. Soul particularly suffers such a fall because of its union with body. But Plotinian falling is not a false way of being in the world; it is rather that arrival in the world itself. For him too, greed for novelty plays a role. It is the Soul's "appetite for things to come" (*ephesis . . . tou mellontos*, III 7, 4, 34) that keeps her in her fallen state. She lives in temporal dispersion, "in a transitional motion from one life phase to the other" (11, 44). It is not clear to me whether existential falling is a necessary, unavoidable, human mode, but hypostatic fall is the very condition of being a Soul. Temporality grounds the possibility for both.

When the Soul falls in the *Enneads*, it falls into the "place" where matter is, for there is no other. Then matter begs to enter it and to spread itself out under the Soul, and it is, recalcitrant though it be, lit up by her. For the body is *in* the soul (I 8, 2, 46). That is Plotinus's figurative account of the origin of place. The ontology is that the temporalized soul is self-distanced and so opens up a place within itself; Plotinus hints that place may arise from time as spatialized before-and-after (III 7, 9, 62). In any case the soul makes the world *within* itself.

Not unsimilarly for Heidegger:

> *Dasein* is essentially "in a world," by reason of the ecstatic-horizontal constitution of that temporality. . . . It "is," with the "outside-of-itself" of the ecstases, "there." If no *Dasein* exists, no world is "there" either. (365)

World is the precondition of space. *Dasein* is spatial only as Care in a fallen mode, that of being cast into the world along with merely extant things. Then "*Dasein* in the literal sense takes in space" (368). Heidegger is purposefully literalizing the German locution *Raum einnehmen*, which usually just means "to take up space." The effect is that *Dasein's* temporality is spatialized and that it "makes room for itself" in which to pursue its possibilities as only a "spiritual" existence can. Merely extended things have no way to be spatial, for space is discoverable only within a world (369) such as only *Dasein* makes for itself.

For both Heidegger and Plotinus, space is thus posterior to and dependent on time (though Heidegger, to be sure, repudiates this thesis in his later essay "Time and Being"). For both it is *within* the temporal being. And, finally, for both it is a place of "lighting": For Plotinus each lower hypostasis is pervasively "lit up," illumined by the next higher one. This primordial light prevents the soul from self-dispersion (V 3, 8, 29 ff.). And for Heidegger, *Dasein* has been brightened or "cleared" (*gelichtet*, 350). The source of the "light" was Care, insofar as it opened *Dasein* to its possibilities. And the ground of Care was ecstatic Temporality. It is what made it possible for *Dasein* to see and take in what is there for it, though *Dasein*, Being-there, is itself this illumination (133), and the light does not come from above.

Perhaps candor requires that I end this parallel account of human temporality on a note of utter difference: Plotinus devotes a whole treatise to showing that happiness is out of time, for it is not a mood or an emotion but rather a fundamental possibility for the Soul, that of being in an undispersed present (I 5, 7, 15), drawn by *eros* deep into thought. *In Being and Time* no word for happiness occurs. Nor for love.

Candor also requires that I conclude these comparable accounts of two grounding enterprises on a note of unresolved perplexity: If the Intellect *is* the Soul before its defection, is not, in a manner of speaking, the Soul the ground of its own apostatic life? If *Dasein* extends itself into the phases of time, is it not similarly the ground of its own ecstatic existence? Is "temporalizing" really a word for the *self*-grounding of time (pp. 152, 196)? But is *causa sui*, self-causation, ever quite intelligible, in any version?

∞ ∞ ∞

What do I think I have shown? Well, first, a remarkable similarity of issues and terms in two philosophers who are about as far apart in their human intention as members of one tradition can be—comparable in the radicalness of their thought, but with the one, Plotinus, caring only about Being *There*, and the other, Heidegger, about Being *Here*. As a mere phenomenon that seems to me an interesting finding. And I think it *is* a phenomenon, because I suspect no direct line of influence at all.

Moreover, an extended inquiry might reveal that the shared urge toward grounding or "fundamental ontology," as Heidegger calls it, entails further deep similarities; examples might be the superseding of discursive thought and the move beyond Being (Rosen, 84, n. 4); some might take pleasure in uncovering a neo-Neoplatonist in Heidegger.

Second, I think I've satisfied myself that when the inquiry concerning time is approached in a certain way, namely ontologically, that is, when the project is not to accept time as an ultimate human condition but to find a ground for it, the outcome will have four deep features: (1) Temporality, (2) Self-distancing, (3) Falling, and (4) Futurality. Here Temporality means a—problematic—primordial or grounding time that shows itself in the temporalizing of a human sort of Being, letting it come to be in some sort of open, stretched, dirempted, self-distanced way that is or can be the undoing of its own Being—a fall or fallenness; such human Being lives in the future, in anticipation.

And third, it seems that time study, whether carried on in a mode mindful or forgetful of the deeper questions, is going to have, by the very fact of possibilities seized or set aside, a moral dimension, by which I mean a relation to the conduct of human life. The moral dimension comes to the fore particularly in the emphasis on the future, which will be taken up in Part Two.

Chapter Four

Augustine and Husserl: The Stretching of the Mind

No two writers on time that I know of are at once farther from and closer to each other than Augustine and Husserl. Their first and last preoccupations are worlds apart, as are their working ways: Augustine thinks only of God; Husserl mentions him but rarely. Augustine seeks to know himself as a creature with a soul and a mind; Husserl investigates himself as a subject with levels of consciousness. Augustine dwells in a created external world; Husserl delves into a self-constituted interior realm. Augustine strains through appearances to existence; Husserl neutralizes existence to find the phenomena.

And yet their thinking about time is so alike that Husserl the Phenomenologist freely acknowledges himself as the avatar of Augustine the theologian. Both seek to explain what are here called the "phases" of time, past-present-future, and both find themselves looking within. Both think of the present, the moment of perception, as pivotal and stretch the mind or consciousness about it. And both find that this "stretching of the mind" leads them to think of time as a two-dimensional continuum, with one axis of temporal lapsing and succession and—so to speak at right angles to it—another axis, one of temporal preservation and contemporaneity.

What brings these two, separated by a millennium and a half of time and a world of thought, so close concerning time? Well, of course, in the first instance the matter itself. They are thinking about the same experience. But second, different though their preoccupations be, they are thinking in the same way, at least about the phases of time. The enigma of human time is not to be referred to a higher ground for resolution but arises and must be discerned *within* the soul or consciousness—of which it may, however, well be the *internal* ground.—Neither Augustine nor Husserl intends to be an ontologist of time.

A. AUGUSTINE

Book XI of Augustine's *Confessions* is his meditation on time, to which Book X, on memory, is the proper prelude. For memory is, partly at least, furnished by the past and partly also by the future. But these are the phases of time apparently inhabited by things no longer or not yet present. It will be Augustine's great discovery that this off-hand impression, that of the two phases extending about the now one is essentially elapsed, and the other unhatched, is false. And he will produce a sketch of the mind (*animus*), almost a diagram, depicting how past and future events are *present* to it without having *presence* before the senses.

I say that this model of the temporal mind is Augustine's "discovery" because he says so himself, and because it is the most readily isolable piece of theory in these two books of the *Confessions*. But Augustine himself says at the very beginning of Book XII that "inquiry has more to say than discovery" (*plus loquitur inquisitio quam inventio,* 1); that is to say, the burden of his thinking is not in the finding but in the asking. So before setting out to give a rendition of Augustine's extractable contribution to the stock of discoveries about time, the one that will be so profitably adopted by Husserl, it makes sense to consider the seating of these two books within the *Confessions*.

The books on memory and time are, as it were, suspended between life and letter. They follow immediately upon the epoch made in Augustine's life by the death of his mother, Monica (Book IX), and they precede his interpretation of Genesis (Book XII).

There is a hymn, composed by his teacher Ambrose, that gives Augustine comfort on the day of his mother's funeral. This hymn is quoted in Book IX and cited in Books X and XI; it sets the tone for the inquiry and will provide its matter. Its beginning is *Deus, creator omnium*: "God, creator of all," and its last two lines are:

> *mentesque fessas allevet*
> *luctuque solvat anxios.*
>
> He eases weary minds
> And relaxes those made anxious by distress.
> (IX 12)

Now it seems to me a fair observation that the question "What is time?" does carry with it a certain intellectual tension, even anxiety. It is evidenced in the sense, alluded to in the beginning of this book, that the question is like no other, but also in a certain tendency to circumvent or explode it. Thus Heidegger accuses Kant of turning away from the ultimate consequences of his discovery, the temporalizing imagination (p. 56), and although Kant might never have agreed to the implications Heidegger attributes to his theory, he can scarcely avoid the charge of stopping short of considering explicitly the transcendental soul's deepest capability, that of bringing thought and time together.

Ludwig Wittgenstein, too, driving Kant's systematic limitation on the competence of human reason to a personal conclusion, says in a passage on Augustine in *The Blue Book* (1933, 26–27) that this "What is . . . ?" question concerning time is, like any other, an expression of "mental discomfort," comparable to the "Why?" question children ask, a question that doesn't necessarily demand a cause or a reason. The question about time, he says, really arises for Augustine in respect to measuring it, and the puzzle comes from a contradiction in the different usages of the word "time": He thinks of it as a length to be measured on a traveling band as it passes us and then applies this process analogously to time itself. The problem, Wittgenstein says, is simple and its difficulty is due to the fascination with and confusion of two similar structures of language: length-language and time-language. Augustine's question is exploded and the pieces are reduced to a grammatical puzzle—grammar understood as a set of strict rules regulating language use. "Philosophy, as we use the word," he concludes, "is a fight against the fascination which forms of expression exert upon us." And, he says, asking for more and better definitions will get us nothing.

Now it is true, as we will see, that Augustine attacks time as a problem by asking how it is that we can measure it, seeing that it is in its forward and backward extension nonpresent, and in its present Now not an extended quantity. But it is quite false that that is the reason why the question of time arises for him, or that he is looking for dispositive definitions when he asks the question "What, then, is time?" Nor is he confused. It might be said that the open-eyed perplexity of one who wants to know is the one state truly exempt from mere confusion, for confusion ends when questioning perplexity takes hold. Moreover, there is really no "mental discomfort" in Augustine's inquiry; the anxiety is rather on the side of him who feels the need to ease anxiety by mortifying wonder and by doing away with the question. But above all, Augustine does not seem to me to labor under a bewitched fascination but rather to rejoice in a serene ardor—though one sensitive biographer judges it to be a somewhat precarious peace (Brown, 178).

It seems to belong to this exposition to say something about Augustine's mode of inquiry and style of writing. The two books in question follow on Augustine's conversion. The distress of assent withheld is resolved; what follows is the desire for familiarity with God and his creation. One might say that Augustine's marveling at the Creation is the counterpart of Socratic wonder—that his faith in God is analogous to Socrates' trust in Being in inducing a stressless tension of inquiry. I mean that neither thinker need be anxious to come to definitive conclusions, to produce strict definitions, complete systems, flattening rebuttals; they rather engage in courting, tentatively yet boldly too, a truth to which they are drawn, never supposing that because a search is endless it has no end. Such questioning is an enjoyment of the object sought (Gilson, 187). For Augustine this frame of mind is rewarded by an extraordinary felicity of phrase. Never was memory or time written of with such suave penetration, such discerning beauty. This beauty is captured in the William Watts version of 1632, close as it is in diction to the King James

Bible. It is the translation reprinted in the *Confessions* of the facing Latin and English Loeb edition, the text from which I have adapted passages here quoted.

Augustine's purpose and end in these books is to enter into a reciprocal relation with his Creator. "Let me know you, my knower, let me know you as I am known" (X 1). Thus, as I said, Augustine's search is not for a God not yet found, but for familiarity with a God fully acknowledged. With his conversion under the fig tree in the garden (VIII 12) and the prayer for his newly buried mother (IX 13) ends the record of that past by which he has been shaped, the "confessions" proper. The books on memory and time accordingly begin with a meditation on "confession" (X 1–4) and the "abyss of human conscience." Augustine seems to be bearing in mind the implicit double sense of *conscientia*, literally "privy-knowledge." Conscience is a knowledge at once deeply private and broadly common, secret and shared. Confession is an exercise of conscience in which both sins and pieties are recalled and communicated to God. To confess is to be open (*manifestus*) before God but silent in the world. So the question arises why Augustine has just published his life and its intimacies in a book. And he justifies himself: It was done to share his joys, to be pitied for his sins, and to stir his readers' hearts from despair.

The two books to come need no such justification. They too are intimately interior, but not in the manner of a private history and an individual life. These are instead the intimacies of the soul we all hold in common (p. xii). They concern the human temporal constitution in its relation to God. The book on memory seems to follow naturally on the biographical recollections just completed, while the book on time begins with an inquiry into the beginning of time that is preparatory to the exegesis of Genesis to follow. But the overwhelming motive for the inquiry into memory and time is, as I said, to know God as he knows us, and that means in our interiority. God is not best loved or learned of in the Creation directly, through the senses, but through the Creation as it reappears within us. Augustine has an exterior and an interior body:

> From which of these is it that I ought to have inquired after God, whom up till now I inquired after through my body from earth to heaven, as far as I was able to send the rays of my eyes as messengers? But the interior [body] is the better one. For to it, as president and judge of the responses given by heaven and earth and everything that is in them . . . all these bodily messengers reported . . . I, the interior man knew all this, I the soul by the senses of my body. (X 6)

So the books to follow will deal, in the service of the inquiry into God, with the interior residue of the senses. Augustine sees this capability of the soul, to have present within it something absent, as the key to memory in particular and to human time in general.

I. Memory

Augustine conceives the inquiry as an ascent by degrees by which he transcends his own nature. First in this journey through the soul, he comes on "the fields and spacious palaces" of memory, where are found "the treasures of innumerable images" of all kinds of things brought in by the senses:

> And yet the things themselves do not enter the memory, but the images of the sensed things are present to me (*praesto*) as thought recalls them. (X 8)

There he is free to call up what images he wants, or to "wipe away with the hand of [his] heart from the face of his remembrance" unwanted memories. There he can sing without voice and see without light:

> This I do inside, in the huge hall of my memory. There heaven and earth and sea with all that I was able to sense in them are present to me, and besides whatever I have forgotten. (8)

The beauty of Augustine's description of the experience of memory derives, of course, from the unabashedly spatial and visual understanding he has of images as being sense-residues, impressions. But this view also disciplines the expansive delight he takes in memory-images.

This fact comes out in a famous letter written in answer to an inquiry from his friend Nebridius in A.D. 389 Augustine here vigorously denies the Plotinian claim that there are images that do not stem from memory and thus from the senses—there are no "phantasms" that originate in the mind itself. But that does not mean that memory is merely coextensive with sense-images. On the contrary, there are things stored for recollection in memory that are not derived from the senses. It follows that there is a nontemporal memory, which indeed appears in Book X.

When Augustine classifies images for Nebridius, it is to claim that all three classes are equally sense-derived: sense-images gotten from the senses, fictive images gotten by addition or subtraction from sense-images, and even mathematical images, which, though apparently produced by reason, are really sensory and error-prone. Augustine, who probably brought the term "imagination" into use, had much more to say about this subject that is not the issue here. Perhaps it is useful to mention that in his book *On the Trinity* (XI 3) he says something further of the process by which sense-memory comes to be. The soul "imbibes" the aspect (*species*) of things, and the retaining memory then informs the "eye of the mind." The imagination is thus entirely an interior analogue to vision. It is vision with a negative sign: vision of aspects in the absence of their bodies. A third power completes this mental "trinity": To memory and mind is added the will, which acts as mediator in focusing the mental organ on the object. Yet these are not three separate powers but three aspects of one three-in-one soul.

So it happens that Augustine must climb way up into memory, to leave its spacious prospects for a narrower but deeper region. These are the "hidden caves" (10) and "remote interiorities" of the memory. These inward places are not properly places anymore (9) since no images are lodged there but the matter itself, such as the liberal arts. These he has learned, but not through any of the five senses, as far as he can tell. There he finds also memory of memory and memory of forgetting (13, 16); this reflexive memory shows that "mind (*animus*) is the very same as memory" (14), for it is amenable to directed thinking and remembering, to deliberate searching for a lost memory.

Three things are in memory: images, arts, and affections of the mind, of which the first only is sense-derived. Yet God has not been found except as a memory, so that Augustine must "go beyond" even his own mind to seek God himself. And he has found him not in the changeable mind nor in any place: "No place—we go to and we go fro—no place" (26). Space and Time have been left way behind with the ascent beyond sense-memory.

Yet the internal world that contains those parts of the Creation that Augustine has experienced in the past, including himself, *is* arrayed in space; "There is present before me heaven and earth and sea and whatever I could sense in them." And in time as well: "There I met with myself; I recall what, when and where I did something and in what way I was affected when I did it"; there also is the future, its events and hopes (8). The elements of the memory book will turn up again in respect to time itself, though somewhat differently delineated. There too will be spatiality, presence, temporal order, and finally, exit from space and time.

II. Time

1. *"In the Beginning"*

Augustine's inquiry into time, the eleventh book of his *Confessions*, begins with his ardent desire to understand the Beginning (*principium*) of the Creation and what Moses said about it. *Da quod amo*: "Give me what I love," he prays (XI 2). "In the beginning" is least of all temporal; above all it means God's word and Wisdom, his Son (7–9). The relation of this Beginning to the beginning of time, the relation of eternity to time, these are the subject of the first thirteen chapters of Book XI. They are meditations on a mystery, in which two clear teachings are embedded: God did not *make* the world, like a human workman, shaping one body from another in accordance with a form before the inner eye, but *created* it in no worldly place nor out of any temporal matter (5). And God did not create the world *in* time, because before the Creation there was no time. In Book XII, Augustine speaks also of two things made by God when time was not yet: pretemporal form and matter, "intellectual heaven" and "invisible and unformed earth" (12, 20). He is offering his own interpretation of Genesis 1, 2, whose events precede the first day (though they are by no means co-eternal with God):

> In the beginning God created heaven and the earth. And the earth was without form and void, and darkness was upon the face of the deep. And the Spirit of God moved upon the face of the waters. (20 ff.)

Evidently there are at least three senses of "Beginning" for Augustine: the eternal unchangeable *atemporal* Creator himself, the changeable though never-changed *pretemporal* constituents of the Creation, and, of course, the *temporal* beginning of our changeable world. And the very fact that there were beings created before time "from which" (*inde*) heaven and earth were made bespeaks the fact that there was no time before Creation, because times only arose when the two aboriginal constituents were joined in the act of creation (29).

There are no times, then, co-eternal with God. And with that, rather abruptly Augustine asks, "For what, then, is time?" (XI 14). The inquiry into the *principium* is dropped for the remainder of the book until the end, and perhaps Wittgenstein's quip about the child that asks why without expecting an answer applies here more grandly: "Inquiry has more to say than discovery"; not all articulated questions can expect an answer. Nonetheless, the chapters to come *do* contain answers, though not about the Beginning.

2. *The Questionable Question*

Augustine begins with a meditation that has come to be the most often cited formulation of the peculiar elusiveness of the question "What, then, is time?"

> Who will have explained it easily and briefly? Who will comprehend it in thought so as to express it in words? And yet, what do we mention more familiarly or knowledgeably in talking than time? And surely we understand it when we talk about it, and also when we hear someone else talking of it. What, then, is time? If nobody asks me, I know; if I want to explain it to someone who asks, I don't know. (14)

What truth has Augustine captured that put his words at the head of so many inquiries into time? What is it about time that makes it so familiar in use and so strange upon reflection? Why is a book by one of the chief scholars of time, J. T. Frazer, so aptly entitled *Time: The Familiar Stranger* (1987)?

The familiarity in use is not, I think, so deep a problem, and I have been arguing in this book—and was in very good company doing it—that talk of time is very often talk of relative motions. But the strangeness in reflection is ineradicable because it ultimately touches upon the position of human Being in the world. So much the more is a crisp formulation of our speculative unease about time needed, and Augustine presents it right away: The future is not yet, the present passes, the past is gone. "How can we claim that that is, whose cause (*causa*) of being is that it will not be?" (14). This way of phrasing the question, in forestalling the prior question "Is there Time?," reveals its own disturbing iridescence. Its very hypothesis, that time has a "What is it?" to be asked about, that is to say, that it has Being, seems to invite the observation that it "tends not to be" (*tendit non esse*, 14). So time is neither a stable being nor outright nonbeing, nor again, a becoming, since Augustine will argue that time precedes all motion of bodies (24), and becoming belongs to bodies. Augustine expresses the special difficulty of asking about a mystery that is known to us in every detail by a confession: "Woe is me, that do not at least know what I do not know!" (25).

In the face of these perplexities Augustine immediately makes an assertion:

> Yet I confidently affirm that I know this; that if nothing was passing, there would be no past time (*praeteritum tempus*); if nothing was coming, there would be no future time (*futurum tempus*), and if nothing was, there would be no present time (*praesens tempus*, 14).

And later on he says plainly: "There are therefore things past and things to come" (17).

The great significance of this fundamental claim is, it seems to me, that it firmly posits external time. Things, events in the world, are "in" time: they come, are, and pass. World-time begins with the first day of Creation and is established by the subsequent six days. There is no time before the created world, nor do the motions of Creation themselves constitute time. But there is a temporal progress, marked from the first by diurnal measures, that, as Augustine is told "in his inner ear," turns for us from expectation to sight to memory. God is unchanging and eternal, he wills *semel et simul et semper*, "once for all and all at once and always" (XII 15)—eternity in a formula—but his Creation contains passage.

It is in the spirit of the *Confessions* that Augustine dwells on the Beginning and not on the End, but it is clear that time that begins also ends and so is finite. Moreover, it is also implicit in the Creation that human time, the time on which Augustine will now meditate, is partial time, time within time, since man was not made until the sixth day and lives briefly and learns of the Creation only in fragments.

3. *Creation-Time*

Thus Creation-time will be a sort of base time or horizontal coordinate, on which human time will be, as it were, erected ordinatewise. Later on, as I have mentioned, Augustine will strenuously deny that this "objective" time is bodily motion (23–24). For differently paced motions as well as rest are to be measured and cannot therefore be themselves the measure. Particularly, he takes issue with the Timaean theory that the motions of the heavenly bodies *are* time. They are, to be sure, signs—signs of days, seasons, and years—but the whirling of the heavens is no more time than that of the potter's wheel, since their motion might be uneven or intermittent: Did not by prayer Joshua make the sun stand still?

Augustine desires to know "the force and nature of time by which we measure the motion of bodies," the "delay itself" (*mora ipsa*) in which the motion, swift or slow, occurs. He seems to be positing an absolute, equable duration, perhaps a prototype of Newtonian flux (p. 153). But it is not this world-time that Augustine pursues, and, in truth, he leaves the matter of physical time dark. In particular, he does not here explicitly connect the mutability and variability of Creation (4) to its temporality, though he does so in the *City of God*, where he asserts that there is no time without some creaturely change (XI 6)—the meditations of the *Confessions* are more radical.

There is one more thing to be said about Creation-time. It is not at every instant starkly divided, as is secular time, into an indeterminate future and a fixed past. Since God knows all times altogether in the eternal Now where all the phases are copresent, just as nothing past is lost, so the future is in principle knowable—though Augustine does not pretend to be able to discover how prophets are instructed. Human imagining and conjecture is a different matter; that he will explain (17–19).

4. *Measuring Time*

Time tends not to be, and yet we speak of it as an extensive magnitude. But where is that length, short or long, that we do in fact sense and attribute to time (15–16, 21–23, 26–28)?

Of time past we must say that it has been long or short, and of time future that it will be so, but of neither that it *is* long or short. If, however, we then say that "it was long when it was present time" we run into the difficulty of determining the present. The present is not the year now going on, nor a month nor the "space of one day" of it, nor an hour. "If something belonging to time is conceived that can be divided into no parts, not even the most minute of moments" (*minutissimas momentorum*, 15)—that would be the present. It would have no "delaylet" (*morula*) and so no length. So "the present has no space."

Augustine frequently speaks of lengths of time in spatial terms. Though he mentions the practice only briefly (26), it is plain from this language that the space-analogue sets the problem as it will dictate the answer. For in space measuring offers no immediate difficulties because the length to be measured is at every moment there simultaneously with the measuring stick and the measuring agency; the measuring length can lie alongside the length measured and measuring observers can lay off the measure at their leisure. Space offers the "delays" that make it possible to complete a measurement. But time runs continuously off into nonbeing. Moreover it does so in one dimension, so that there appears to be no measuring stick of time to apply to or to lay "alongside" of time. You cannot approach time from the outside, so to speak—not if you deny, as Augustine has, that time is a motion or is primarily measured by motion.

Two points are plain to Augustine: that we do measure time, and that we measure times as they are passing (*praetereuntia*, 21). He knows the first beyond a doubt: "And so, my God, I do measure, but what I measure, I do not know" (26); especially in scanning and reciting verse is he aware of measuring long and short lengths. That time must be measured in its passing follows from the sensory character of what is measured when time is measured, say a verse line of iambic feet: "D̆ēŭs Crĕātŏr ōmnĭūm," eight syllables alternately short and long. It is that sensing that is of the present and is passing away (16); in our language, the sense of time passing, much or little, is intimately linked to perception.

This is a circumstance that Augustine as a musician and lover of sacred poetry is acutely aware of: The present of perception is the font and origin of time.

5. *The Pivotal Present*

If, or whenever, or wherever, anything *is*, it is present and in the present (17, 18, 20). "If I am not up [to knowing where past and future things are], yet I know this: wherever they are, they are not there future or past, but present things" (18).

And now comes the first of a series of discoveries. He begins with the observation that in spite of the present alone having being, what he has heard since he was a boy, that there are "three times," past, present, future, must be true. For people do see and tell of past and future events, and what is not somehow present in time cannot be visible (17)—Augustine's positive converse to his claim that what is visible is present. Hence we may speak properly of three times, though not as past, present, and future. Instead we must speak of

> a present (*praesens*) of bygones, a present of presences (*praesentibus*), a present of future things. (20)

In other words, all events are temporally in the present. This ingenious resolution has in it two perplexities. One is that the present is not a stretch of time or space, and "no-space we do not measure" (21). This difficulty, so much agitated later as a problem of uniting conception and perception, the difficulty of the "specious present," will be resolved conceptually only partially (next section). But it will be justified theologically quite completely, for I shall show below that the moment of perception is for Augustine analogous to the eternal Now of God's omniscience (p. 124).

The other perplexity is that there are, by this solution, events and things that are, so to speak, adjectivally temporal, "things futural," "things bygone" (*futura, praeterita*), but that no longer really belong to their proper temporal phase. This problem Augustine will now solve.

6. *The Stretching of the Mind*

The passage just quoted immediately continues:

> For some such there are in the soul, and I do not see them elsewhere. The present of things gone by is memory, the present of things present is sight (*contuitus*), the present of things future is expectation. (20, also 28)

Memory, expectation—these lend presence to the things that have gone by or are yet to come before perception, from which they diverge on either side, so to speak.

What is in memory (and in expectation and foresight as well) is, of course, not the things themselves but their images; the things, in "going by the senses, have fixed, as it were, traces (*vestigia*)" of themselves that can be expressed in words (18). We can now see why the book on memory, on the internal world that is visible without being present, was a necessary prelude to the book on time.

In the essay *On Music* Augustine makes the connection between sight and memory. *Contuitus* in the passage above, taken literally, means "together-sight," surveying contemplation. Thus in *On Music* he says that just as the eye captures and unifies the "spaces of place" (*spatia locorum*), so memory, which is "a sort of" (*quasi*) light for temporal spaces (*temporalium spatiorum*), captures the disparate moments as they push each other along and makes them, as it were, one simultaneous space of time (VI 8, 21). The quasi-light of memory gives its contents a spacelike copresence: "Yet what do we measure if not time in some space?" (*Confessions* XI 21).

Thus the events that Creation-time brings forth are taken into the soul, into memory (and whatever goes for memory, applies *mutatis mutandis* to expectation, which is just future-directed memory), and there they preserve their presence continuously. This notion approaches a resolution of the "delay" question concerning perception. The moment of sensation is just the point of intake, from which the ever-present past and future stretch before and behind, analogously to the extended segments of a straight line disposed simultaneously about a pivotal point. No perception of temporal wholes like melodies, Augustine says in *On Music*, could ever arise if past events were not thus attached to and present in the moment.

Not motion measures time but *I* measure it, Augustine had said before. And what he measures are the perceptions that have passed: Nothing is bygone if no thing has gone by. Bygones are measurable because all times are copresent, since memory and expectation are in the present. Therefore time is somehow like space. And now he draws his great conclusion from these propositions, the one for which his mind was aflame, the one "whose discovery is new" (*nova est inventio eorum*, 22). He begins by observing that a song, a foot, a syllable, a whole life (28), are so measured, by means of a spacelike copresence:

> Whence it seems to me that time is nothing but a stretching (*distentio*). But of what thing I do not know, and I marvel if it be not of the mind itself. (*ipsius animi*, 26)
>
> It is in you, my mind, that I measure my times. Do not impede me, that is, do not impede yourself with the throngs of your affections. In you, I say, I measure times. The affection that things gone by produce in you remains even when they have gone by; that itself do I measure as present, not those things that have gone by so that this affection might be produced. (27)

And Augustine immediately mentions silence, absence of sensory stimulation, that stumbling block of those who, like Aristotle, connect consciousness of time to external motion: We measure a space of silence when "we stretch (*tendimus)* our thinking" to be the measure of a voice "as if" (*quasi*) it now sounded. Thus our "present attention (*intentio*) puts the future over into the past."

A short observation on all the "stretching" words employed by Augustine is in order. *Attentio*, which Watts well translates as "attentive marking," is the mode of the mind in touch with the perceptible world. *Intentio* is used in the same way; the mental mode indicated by this term, descriptive as it is of the mind's stretching toward an object, becomes as "intentionality" a key word for the Phenomenologists and in particular for Husserl's treatment of time (p. 129). I should add that Thomas Aquinas, in the long discussion of intention (*Summa Theologiae* II 1, Question 12), points out right away that intention belongs first of all to the will—recall that for Augustine the will brings the other faculties together in the act of knowing. Now this willful stretching of the mind *toward* an object implies the stretching of the mind *in itself* (Ricoeur 1984, 1:20).

Distentio is that stretching of the mind, which is Augustine's chief discovery. "Distension" is close to "dimension"; *di* (which become *dis* before t) signifies what comes apart, as in the Greek term *diastasis* that figures so importantly in Plotinus's account of the temporal soul (p. 99). I think that Augustine chose the tending *-tentio* nouns over the static *sta* words that dominate Plotinus's (and Heidegger's) time-language because of the longing and striving that characterize his search for time and temporality itself, a yearning not distinguishable from his desire to be close to God. The difference in language in fact illustrates what I meant in the beginning by saying that Augustine is not an ontologist of time: He departs from his master Plotinus, in whose *Enneads* he had long been absorbed, by thinking of time, and the fall it implies, not as the self-distancing of an intellectual ground but as the double-directed stretching of the created soul, which as it stores the passages of the world also tends away from them.

And so we come to a final tending or stretching word, *extentio*. This noun does not actually occur but the adjective does, *extentus*, as does its verb, *extendo*. It is Augustine's word for his striving to leave the dimensions of time and motion (29), as we will see. And with its introduction, *distentio* changes its meaning. It is now "distraction."

7. A Putative Diagram

I cannot really pretend to think that Augustine ever drew the diagram shown here. Yet it seems to me justified by everything he has said. First of all, we must suppose that Creation-time flows on, not as a part of God's eternity but as a mysterious finite fall-out from it (p. 118). In it occur those events or things—Augustine characterizes them simply by a neuter plural "the bygones" (*praeterita*)—that can be and have been taken in by the senses. This external time may be thought of as flowing linearly from past to future, from the beginning, the act of Creation, to the end, the Last Day and Judgment told of in the *City of God* (XX); this flow may be represented by a static (finite) line, but is to be conceived as a steady passage, a rolling band, so to speak, moving forward toward the end of time (as it is depicted here), or alternately as sliding backward under pressure of the foreordained future (*Confessions* XI 11).

Some part of this event-flux is taken up by the senses and conveyed into the mind. The mind itself is dimensionally stretched, and is therefore also fairly repre-

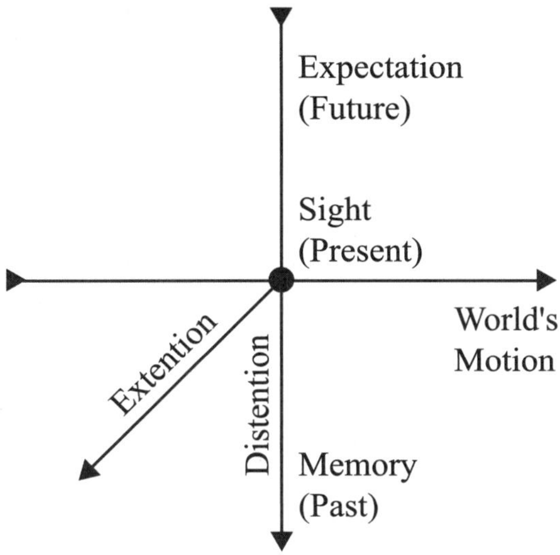

sented as a line, a line that sits—why not orthogonally?—astride the external time-band that passes under its noting attention. Mind and world, mental time and natural time, phase-time and succession-time, intersect at the origin of the coordinates, for that is where presence takes place. Here the world with its places and events is *now*, and the things *themselves* are present. As external events are taken up into memory and as expectations of the future are realized, the vertically stretched mind is filled in the order of perception and realization. Perhaps the line may be thought of as stretching further, as being "distended," with the accumulation of memories until death. The purely temporal order is linear, but we know that memory, where the things that are no longer there themselves are lodged, also has dimensions and space enough not only to accommodate their individual spatial, that is, their visual forms, but also to see various past things panoramically, together.

Just how the affections produced in the mind by passing events drop down into memory and how images of the realistically expected future eventually come before sight as perceptions—these are questions Augustine no longer pursues. But it is just these aspects of time-consciousness that Husserl will take up in detail and, moreover, incorporate in his famous time-diagram, which is an elaboration of Augustine's undrawn figure (p. 135).

8. *Before the Beginning*

I have added a third dimension, the *ex*-tention mentioned above. For when all is said and done, the stretching (*distentio*) that makes the mind temporal brings a tension and "distraction" (named *distentio* as well) into Augustine's life, which is thereby drawn "many ways by many things" (29). Here he prays to Christ, the Son of man, who is the mediator between God who is one, and us who are many. He prays that he may be gathered away from the old life and forget what has gone by:

> not stretched apart (*distentus*) into things in the future or things passed away, but stretched forth (*extentus*), not according to distraction (*distentio*) but according to attentive thought (*intentio*). . . . (29)

Augustine is, it seems, playing on two meanings of *distentus*, the one used up till now, from *distendo*, "to stretch apart," and the other probably from *distinio*, "to divert attention, to distract." In any case, this stretching of the mind is to be gathered in and redirected into a stretching-forth, out of time into the contemplation of eternal delights neither yet to come nor ever to pass away. He himself "will stand and grow firm" in God *(Et stabo atque solidabor in te*, 30). And he urges those who ask what God did before he made heaven and earth:

> Let them stretch forth (*extendantur*) rather toward those things that are before (*ante sunt*) and understand that you are before all times, the eternal Creator of all times, and no times are co-eternal with you. (30)

So the inquiry culminates in a focused desire to recollect the dis-tention of the mind dilated by time into an ex-tention of the soul out of this world and into eternity and the timeless Beginning.

9. *The Image of Eternity*

Man is made in the image and similitude of God (XIII 22). Does the temporal mind of human beings appear as an image of God's eternal mind? Yes, it does, as the last chapter of the book on time intimates.

God does not, to be sure, know the Creation in our mode, as a singer or listener, whose "mood is varied and whose sense is distended" by expectations and memory of the notes (31). The eternal Creator of minds knows everything together and at once in his mind, the whole creation from the beginning to the end.

So too, the human memory holds together within its "vast and infinite interior . . . high mountains and enormous billows of the sea and broad courses and rivers and stars and the compass of the ocean" (X 88), as well as knowledge not derived from the senses, and above all the affections of its own soul. So the memory contains the world, though not the things themselves but only their images. In a word, it holds an image of the Creation through time.

It is only right to point out that this speculation on the memory as an image-containing image of God's world is quite separate from Augustine's theological employment of memory, which together with understanding and will forms one of the trinities, mentioned before, by which the "inner man" images God's three-in-one nature (*On the Trinity* X–XI; *Confessions* XIII 11). But my speculation is not unfounded. Augustine tells us that when God surveys his Creation in parts he sees it as merely *good*, but when he surveys it all at once, he sees it as *very good* (XIII 28). Now in the very last chapter of the whole *Confessions* Augustine says:

> We therefore see these things that you have made because they are; but they are, because you see them. And we see outside that they *are*, but inside, that they are *good.* (XIII 38, my italics)

Thus the outer and the inner vision of human beings is assimilated to God's survey of the Creation—reduced by one level, as it were. God sees it as good in parts and as very good as a whole; we see it externally as merely being and internally as being good.

It is, moreover, in the book on time that Augustine discovers—though again not explicitly—the human *temporal* mind as an unmistakable image of God's *eternal* mind. For all things are copresent in eternity, and that is what Augustine's "new discovery" effects for the human mind: the copresence of the three phases of time. I might put it this way: The "distended" human mind, as it rides athwart the temporal world taking it in through the senses, is able by reason of its stretching to gather its part of temporal Creation back into one moment, the present. But that simultaneity of all times is the image—like all images, reduced in being and in compass—of the replete *standing Now* of God's eternity.

∞ ∞ ∞

Here, finally, is a summary of the results that Augustine's inquiry may be said to have yielded for the purely secular study of time, results that will be recaptured and elaborated by Husserl.

(1) The human mind is temporally stretched: It has a certain linear dimension in which past events are retained in memory and future events foreseen in expectation, in the order of their eventuation.

(2) External time, Creation time, is a linear passage: The world in its changes has a temporal order that, as it flows, carries things that are coming into things that have gone by, or the reverse.

(3) These two time-axes are orthogonal to each other: Human time and world time, the time of the soul and the time of things, are imcomposably far from having the same direction, or mode of passage.

(4) The two time lines intersect in a point: World and mind are at one in each moment of perception, when the things themselves confront sense and are noted by the mind; because of the world's passage, such moments follow each other continually.

(5) The phases are copresent in the mind: Memories of past and expectations of the future events are erected, so to speak, upon the point of perception and are bound to it and its situations in world-time.

(6) The sense of time depends on this copresence: Without present memory, time would be continually lapsing and could not be said to have a length; the dimensional mind accommodates stretches of time in the present.

(7) The phases of time arise only in the temporally constituted mind: Without memory and expectation there is no past or future.

(8) Creation-time is not derived from its *principium*, its Beginning, ontologically as from a ground; that is to say, it is not educed rationally as a lesser from a greater being. Creation is left as a mystery.

B. HUSSERL

Augustine is the first writer on time to be named, praised, and quoted by Husserl in the core text of his analysis of *time-consciousness*, that "age-old crux of descriptive psychology and theory of knowledge." Augustine was the

> first who felt deeply the tremendous difficulties that lie here and who toiled over them almost to despair. . . . Chapter 14–28 of Book XI of the *Confessions* must even nowadays be studied thoroughly by everyone who occupies himself with the problem of time. For knowledge-proud modernity has not advanced so gloriously far or advanced so considerably farther in these matters than this great and earnestly struggling thinker. (Introduction to *On the Phenomenology of Internal Time-Consciousness*)

Now the affinity of approach will be considerable, but not, as I have intimated in the last chapter, the likeness of mood (p. 113). Augustine's inquiry into time is a devotional enjoyment; because he firmly believes, he wants to understand. Husserl, on the other hand, secular and subjective, is engaged in never-ending battle with the most slippery of possible grounds for the self. His endless and ever-revised cogitations are, humanly speaking, a different enterprise from Augustine's serenely provisional musings. Yet the felt congeniality seems, to me at least, to betoken a real similarity: the kindred faith of those who never suppose that just because the search is without end for us, therefore it has no end in itself. I make this observation here because it will function as an incitement to interpretation: It makes Husserl's multiple self-emendations tolerable. The numerous detailed similarities in the understanding of time itself will be noted in their place.

I. The Perennial Question

Naturally Husserl ends the first paragraph of his book on time with the quotation of quotations on the subject: *si nemo a me quaerat, scio; si quaerenti explicare velim, nescio*—"If nobody asks me, I know; if I want to explain it to him who asks, I don't know" (p. 117).

It is profitable to try over and over to make out wherein lies the elusiveness of a question whose answer one knows only when not stirred to give it. No writer on time will bring the questionableness of the question, why the explicit knowledge of time is so fugitive when the lived knowledge is so slippery-smooth, under closer scrutiny than Husserl. His main means will be the analysis of time into levels, three in fact, the deepest and most problematic of which will possibly be human Being itself—a tricky problem taken up here in the next to last section (p. 150). It occupied Husserl, as I said, from early to late: In the introductory remarks to the lectures that were to form the core of his book on time Husserl said of his early work, the *Logical Investigations*, that "in it all the problems of a *Phenomenology of originary time-intuition* were, so to speak, silenced to death. The extraordinary difficulties that lie here, perhaps the greatest in the whole of Phenomenology, I was not able to master . . ." (*Husserliana* X, xvi).

This is a good moment to bring together a few of the quotations that form, one might say, a plaintive tradition concerning the peculiarities of the study of time. Augustine is followed by Simplicius in the mid-sixth century, who puts the sentiment in Greek:

> The existence of time is very clear not only to the wise but to everybody. But if asked whatever time is, even they could scarcely answer. (*Commentary on Aristotle's Physics*, 695)

In 1905, Husserl goes on in the next paragraph after quoting Augustine:

> Naturally, what time is we all know. It is what is most familiar. But as soon as we attempt to give ourselves an account . . . we become involved in the strangest difficulties, contradictions, confusions. (*On the Phenomenology of Internal Time-Consciousness*)

And finally, later in our century, a poet, Paul Valéry, writes:

> I stop the word *Time* in its flight. This word was utterly limpid, precise, honest, and faithful in its service as long as it was part of a remark and was uttered by someone who wished to say something. But here it is isolated, caught on the wing. It takes revenge. It makes us believe that it has more meanings than uses. It was only a *means*, and it has become an end, the object of a terrible philosophical desire. It turns into an enigma, an abyss, a torment of thought. . . . ("Poetry and Abstract Thought," *The Art of Poetry*)

One reason why Husserl will go so far with his analysis of time is that his method of inquiry allows him to focus this "terrible philosophical desire" specifically on the familiar features of time and to reach deeper levels *through* them.

II. The Phenomenology of Internal Time-Consciousness

A good way to begin working oneself into Husserl's method and its results is, it seems to me, to attend in turn to the terms of the title of his book on time. Unfortunately it is, in this case, impossible to go to the text immediately; a minimum of text history is necessary because of the complicated circumstances of publication that dog this as most of Husserl's work.

Edmund Husserl's most accessible book on time, the one on which the explications below will largely be based, is called, as in the heading above, *On the Phenomenology of Internal Time-Consciousness* (*Zur Phänomenologie des inneren Zeitbewusstseins*; this, as all other works by Husserl, will be cited here by the paragraphs that have become standard). The first part consists of lectures delivered at the University of Göttingen in 1904–05. The second part consists of appendices keyed to those paragraphs. The text is the fruit of Edith Stein's heroically laborious "working-up" (*Ausarbeitung*) of daunting stacks of manuscripts and notations. It was carefully inspected by Heidegger, who assumed the title of editor, and was published in 1928. An English translation by James S. Churchill was published by Indiana University Press in 1964. (Since Husserl's German is occasionally somewhat more vivid than this English version I have sometimes altered it.) The 1928 publication contains Husserl's ever-changing thinking from the time of the lectures through 1917, mingled together in a sometimes confusing, though often handy way.

The German text is printed in *Husserliana*: *Edmund Husserl, Gesammelte Werke, X: Zur Phaenomenologie des inneren Zeitbewusstseins, 1893–1917* (The Hague: Nijhoff, 1966), edited by Rudolf Boehm. This invaluable volume, with its illuminating editorial introduction, contains, in nearly five hundred pages, the preparatory studies, reworkings, emendations, notations and, above all for present purposes, the diagrams that make the reading of Husserl so complicated a business. It was translated by John Brough (Husserl 1991).

It belongs intimately to Husserl's treatment that his inquiry, though directed toward a settled solution worthy of a strict science about a subject he conceived of as "a completely *self-contained sphere of problems* and one of exceptional difficulty" (*Ideas*, 1913, para. 81), is always a work in progress. He is always asking himself if what he has clued out can be correct, pursuing with unremitting meticulousness subtle difficulties raised by proposed solutions, spinning out with obsessive persistence "immensely intricate analyses, which burgeon and blossom like a tropical forest" (Findlay, 13).

Let me now pass the terms of the title in review, as a way into Husserl's analysis.

(a) "Phenomenology" in general betokens a "new, descriptive, philosophical method" ("Phenomenology," *Encyclopedia Britannica*, 14th Edition, 1929; the article, by Husserl himself, is the main reference below). The term, when applied to particular fields of investigation, is used with the definite article: "On *the* Phenomenology of Internal Time-Consciousness."

The word, which is derived from the Greek *phainomenon* (what appears) and *logos* (an account) has had a wide range of uses in philosophy, from Kant's narrow application to the route by which the mere appearances of motion become objects of experience (*Metaphysical Foundations of Natural Science,* 554) to Hegel's grand designation for the story in which Spirit appears through a gallery of figures in time (*Phenomenology of the Spirit*). Husserl now appropriates Phenomenology as the proper name for a method of inquiry whose three chief features are indeed focused on appearances.

First, Phenomenology is concerned with the true locus of appearance, our own consciousness. The world and we ourselves appear to ourselves "in consciousness." Conversely, what we are not conscious of doesn't appear. Therefore Phenomenology attends to the "psychical 'acts of lived experience' in which [the phenomena] are apprehended. . . . These 'appearances' are phenomena, whose nature it is to be a 'consciousness of' their object, real or unreal as it be."

The word "phenomenon" is thus used in a double sense: act of consciousness and content of consciousness (*Husserliana* X, 336). Consciousness belongs to a subject, but Husserl is not making the world subjective in the ordinary sense of making it relative to us; on the contrary, he is trying to find a radically suppositionless founding for it.

Thus arises the second feature: Every appearance, as we commonly suppose, is an appearance of something, and so—it is no coincidence—is every consciousness a consciousness *of* something. It is the nature of our human awareness to be able to distinguish between its acts and its objects by a "reflection" in which we disregard the matters and values of the objects so as to focus on the subjective experience. Husserl learned from his teacher Franz Brentano to use the scholastic term "intentionality" for the essential reference-character of consciousness. One way to see the relation of intentionality to the phenomena is this: Certain philosophers, particularly Plato, have thought that the world was all appearance, but appearance *of* the true Being behind it; Husserl thinks that consciousness harbors appearance, but appearance *of* objects, real *or* unreal.

Intentionality is, however, more than just a name for the relationality of consciousness. It also preserves something of the usual sense of "intention," insofar as acts of intentionality confer meaning on their objects, even "interpret" them. Intentionality is always, somehow, self-consciousness—the self is involved as willing or comprehending.

And so the third feature is introduced: the indifference to the existence or nonexistence of these objects. Besides the strenuous self-inspection clearly required by the Phenomenological method, its "reflection" requires a special discipline, variously referred to as "reduction," "switching-off," "bracketing," and *epoche* (Greek for "holding off"). What is held off is existence: "Our comprehensive *epoche* puts, as we say, the world between brackets, excludes the world which is simply there." In concentrating on its phenomena, the being or nonbeing, existence or nonexistence of the objects intended is set aside in order to allow a pure focus on intentional experience and its structures, undistracted by existence questions. Husserl is deliberately contravening what he calls the "natural attitude," which presupposes the world's existence. In the case of time, the phenomena surviving the existential reduction are "internal."

I should add that besides the Phenomenological reduction, the "bracketing of being" just described, Husserl speaks of two others: the reduction of the cultural world to our lived world, and an ultimate reduction that leads from the internal consciousness of appearances to the I, the self or subject that underlies them; this last reduction is called in Kantian language "transcendental" (Kockelmans, 31); here will be located the ultimate questions about time.

Accordingly, the lectures on time begin directly with the switching-off or exclusion (*Ausschaltung*) of "objective time" (1). The existence of world-time, the time of nature and natural science, of concrete duration, and of clock-time is set aside (*Husserliana* X, 339). Husserl illustrates the procedure with an example from space: We can make all sorts of observations about spatial appearances such as juxtapositions, superpositions, interpenetrations, but these are not necessarily objective spatial relations, for we would not say, for example, that a partly occluded object is only partly there (1, and 9, n. 7). So too, we can go far in noting temporal phenomena without making inferences about the being of time.

The cultural reduction is not a focus of this particular Phenomenology, that of time, while, as I said, the transcendental reduction finally comes on the scene as a huge and remarkable question mark (p. 152).

In withdrawing from the objective or external world and its time we come to concentrate on another time: immanent time and its existence (1).

(b) "Internal" means, in Phenomenological contexts, "immanent," that is, what stays within the mind (8, 9). Husserl sets out the pair of concepts "immanence—transcendence" in the *Idea of Phenomenology* (1907, 5). Refinements aside (and nothing Husserl expounds is without them), the pair is construed to turn the normal approach upside down: Transcendent is that which we ordinarily call the real and the reliable; it is what goes beyond the internal, and this realm is for Husserl doubtful—doubtful in the sense in which Descartes made a method of doubting the existence of the world. Immanence, internality, on the other hand, is without doubt; immanent objects are directly given to internal sight, to self-inspection. Husserl ends the last work published during his lifetime, the *Cartesian Meditations*, with a quotation from Augustine's *On the True Religion*: "Do not wish to go outside; go back into yourself. Truth dwells in the inner man" (39).

On the scene of immanence many of Husserl's most subtle and often-rethought distinctions appear, namely those having to do with the very character of cognition. Only one of these is indispensable for understanding Husserl's theory of time. It is what he calls "the schema: content of apprehension–apprehension" (1, n. 2). In the note cited, Husserl is beginning to disavow the schema and thereby hangs a tale (*Husserliana* X, xxxv ff.; p. 146). What is meant is a fine distinction within consciousness, already set out in the *Logical Investigations* (1900, Fifth Investigation, para. 14), apparently called for by the combination of immanence with intentionality. Consciousness has a kind of immanent content, a pre-intentional sensed raw material, and the intentional act of apprehension (*Auffassung*) actively constitutes an object of perception through or by means of this sensed matter; it animates it by various intentional acts. In short, we have various mere sensations, such as sounds, and we make something of them through acts of interpretation; for example, the sound material is construed or constituted into an element of a musical composition (Brough 1987, 11).

This schema will, as I said, later be dropped. But the notion of "constitution" remains, for "it is the crucial characteristic of subjectivity" (Kohák, 123 ff.).

(c) "Time-Consciousness" encompasses both the particular acts of consciousness—acts of intentionality, of course—and the immanent temporal objects, the "lived experiences of time" (*Zeiterlebnisse*, 1), which are constituted by these acts. The analysis will have to recover elements that have no separable existence, and this fact will make for a somewhat embarrassed description. Ultimately it will go beyond the observation of Phenomenal time and its particular constituting acts; it will become a transcendental inquiry and touch on human subjectivity itself.

To begin with, however, it is consonant with the character of Phenomenology that this particular inquiry begins with time as a given experience. Thus the question of the existence of time cannot arise. Husserl never expresses himself on this matter, but it seems as if even before the Phenomenological "switching-off" by which world-time is excluded, the whole question of the Being of time had been excluded.

The justification for this exclusion might be said to be found in the claim that there are self-evident a priori laws of time such as are usually gotten from the mathematical model: The fixed temporal order is an infinite series, in which no two times can be simultaneous; this order is transitive, "so that to each time belongs an earlier time and a later one" (2). Of course, if the laws of time are a priori, so is time. At any rate, two issues will not arise in the analysis of time-consciousness. Neither the existence of time, nor its definition, will be at issue.

III. Predecessors

Husserl was a man willing to express generous gratitude. Three names that he himself mentions are of chief interest here.

(a) Augustine is, of course, *the* great predecessor of Husserl in the inquiry into time. In this case, influence might really be said to be at work, since Husserl found in the tenth and eleventh book of Augustine's *Confessions* the same preoccupations with memory and time that he would so searchingly elaborate and modify. Let me just name them here and leave the detailed comparison to the next section on Husserl's time diagram (p. 138).

First, of course, is the analysis represented in those diagrams, which I have called the "orthogonal" view of time because it requires time to be represented as split into two mutually perpendicular axes. It is Husserl's guiding idea. Whether he got it from Augustine or from William James (p. 133) or merely found fellow thinkers in these precursors, I do not know, since Husserl does not say—probably the last, since he is usually explicit in crediting through detailed critique those who open the way for him with specific notions.

Besides the fundamental idea of "orthogonal" time, Husserl might have recognized in the *Confessions* faint intimations of that penumbral or "specious" present that constitutes his second great discovery, that of "retention" and "protention." Moreover, he might have found in the *Confessions* the word *intentio* for the act that conveys past to future (p. 122). A fourth Augustinian observation that will become of crucial importance to Husserl is that of memory of memory (X 13), which will, as "retention of retention," function as the bridging connector, so to speak, between the two orthogonal consciousnesses (as will be explained below, p. 143). A fifth interest in which Augustine precedes Husserl is in the image-character of memory (X 15); here Husserl will add the entirely novel modification of "presentification" (p. 146). And finally, just as Augustine adds an "extention," a dimension of timelessness representing a return to God's present (XI 29), so Husserl will come to an atemporal aspect of time-consciousness in the "absolute constitutive flux" of the atemporal subject (34).

This coincidence of preoccupations and parallelism of solutions becomes, it seems to me, the more remarkable the better the deeper differences between the two inquirers into time are kept in mind. Here are the chief of those differences: Augustine never lets go of the existence of God and his world; these are, of course, "bracketed" by Husserl and do not enter into his analysis of time, though they do lurk behind it; he says, most remarkably: "God's infinite consciousness comprehends all time 'at once.' This infinite consciousness is atemporal. . . . The divine consciousness is the ideal correlate of objective time and of the objective world and world-development" (*Husserliana* X, 175). Furthermore, Augustine searches into the "inner man," his soul and his mind (*anima* and *animus*), while Husserl deals with a "transcendental subject" and its consciousness. But although these two versions of human Being are separated by the divide between Christian antiquity and secular modernity, even here a deep affinity appears: Both Augustine and Husserl proceed by means of meticulous introspection; the quotation that stands at the end of the *Cartesian Meditations* is Husserl's acknowledgement of this affinity: "go back into yourself. . . ." And neither ever gets finished; Augustine's saying in Book XII could be Husserl's:

> Inquiry has more to say than discovery, and the seeking is lengthier than the obtaining, and the hand that knocks is more laborious than that which receives. (1)

(b) William James is acknowledged by Husserl as the source of "strong impulses" in the introductory remarks to the lecture series of which the Phenomenology of time was a part (*Husserliana* X, xvi). A remarkable paragraph in James's *Psychology* of 1892 (James's own abridgment of his *Principles of Psychology*) shows that the impulse was *very* strong indeed. In the chapter on "The Sense of Time" under the heading "The feeling of past time is a present feeling," James says:

> If we represent the actual time-stream of our thinking by an horizontal line, the thought *of* the stream or any segment of its length, past, present, or to come, might be figured in a perpendicular raised upon the horizontal at a certain point. The length

> of this perpendicular stands for a certain object or content, which in this case is the time thought of at the actual moment of the stream upon which the perpendicular is raised.
>
> There is thus a sort of *perspective projection* of past objects upon present consciousness, similar to that of wide landscapes upon a camera-screen. (285)

What is described here but the orthogonal time-diagram of Augustine with the addition of Husserl's oblique lines (the projecting rays in James's terms; p. 136)? I might mention here that a similar diagram was produced by C. D. Broad in 1923 (Gallagher, 38).—That makes three fairly independent renditions of the "orthogonal theory" of time.

But the diagram is not the only prefiguring thought to be found in James's two brief chapters on time and memory, which contain more succinct insights—insights Husserl will elaborate—than any other modern treatise on time I know of: the idea of the "specious present" (James invented the term) as the "saddle-back of time" (280); the insight that "*A succession of feelings, in and of itself, is not a feeling of succession*," so that the latter must be treated as an independent fact requiring special elucidation (285); the notion of "symbolic" long-term conceptions that, unaccompanied by a sense of duration, are known only by dates (283) and that Husserl will call "empty" because no intuition is given in them; the intimate involvement of time, self, and object in memory; the positively Husserlian-sounding definition of the past by its "phenomenal contents, an event imagined as located therein, and owned as part of my experience" (288); the distinction, later elaborated by Husserl in the same terms, between "retention" and "reproduction" (288).

James also makes many additional observations on the varying speeds of inner time-experience (284–85), a topic that Husserl touches only lightly but that becomes a principle of composition for Thomas Mann's *Magic Mountain*: the shortness of experience-filled time as it passes and its expansiveness in retrospect; and again the dilation of empty time experienced as mere time, time per se, and its brevity in retrospect (Mann, ch. IV, "Excursus on the Sense of Time"). What is of chief importance for Part Two of *this* book is James's admission of a sense of *mere* time, odious and insipid though it may be: "the feeling of bare time is the least stimulating experience we can have" (285). For I will there try to come to grips with "mere" inner time (p. 206).

(c) Franz Brentano is remembered by Husserl in the introductory remarks to the 1904–05 lecture series as "my teacher, gifted with genius," who introduced him to the analytic clarification of problems concerning the comparison of imaginative and perceptual representation (*Husserliana* X, xvi)—problems that will, of course, arise in thinking about the differences between memories and present impressions (p. 145). It was also from Brentano that Husserl learned his own use of the term "intentionality."

Nonetheless it is to a critique of Brentano's conception of the "origin" of time that the first section of the *Phenomenology of Inner Time-Consciousness* is devoted

(3–6). The naturalistic psychologists, Husserl says, who know of no distinction between outer stimulus and inner phenomenon, make the obvious error of confusing the duration of sensation with the sense of duration. This is not the mistake Brentano made (any more than did James). Where he does go wrong is in attributing the production of the past to the "imagination" (*Phantasie*, 3). When a perceived tone—just as for Augustine hymns were the prime example of a temporal event, so for Husserl it was melodies—has been "pushed back," in Brentano's account the imagination immediately reproduces it according to a "law of primordial association," but with a past-character attached to it. Therefore the imagination works, as Brough puts it, like "a psychic machine" continually reproducing perceptual contents with a past (or future) label (Brough 1989). Thus "this imagination proves to be in a peculiar mode productive"—a mode Husserl cannot accept. Moreover, when the imagination furnishes its continual moment-reproductions, these "surrogate phases," newly produced, exist *now*; whence then, setting aside the temporal index (which is a mere attached word), comes the consciousness that an experienced moment is past? How do I know that the experience that the imagination has now newly reproduced is identical with, or at least similar to, the one that has passed?

Husserl concludes by crediting Brentano with being the first to distinguish radically between primary contents and acts of consciousness, a distinction related to that made above, between the duration of a perception and the sense of a duration. But he criticizes Brentano for missing the crucial characteristic of the act of consciousness, this very sense of duration (6). To anticipate Husserl's main resolution of Brentano's difficulty: Duration is *perceived*. Much will depend on this insight.

IV. The Diagram of Time

The diagrams shown here are adaptations from the diagrams Husserl allowed to be presented as the first fruit of his investigations, under the title "The Continua of the Phenomena of Running-off (*Ablaufsphänomene*): The Diagram of Time" (10, also 43). They are compendia of Husserl's chief discoveries about our sense of time, but what is more, they are displays that suggested to him ways of resolving subtle problems inherent in his theory.

The working notes in *Husserliana* X, beginning in 1904, show many variations and reworkings of substantially one and the same idea: abscissa, ordinate, and oblique connecting lines (210, 230, 235, 330, 331, 365, 412). The reasons that it seemed right to produce my own conflations and adaptations were (1) that all the printed diagrams are partial, and (2) that in several diagrams the letters don't quite match the text, the directions are inconsistent, and Husserl himself suggests inverting certain parts of his figure for perspicuity.

The two diagrams, although they look like no one single picture drawn by Husserl, include the representable elements of his theory with rationalized lettering and directions. Their explication will take up this section, while the terms used in the explication will be further considered in section V.

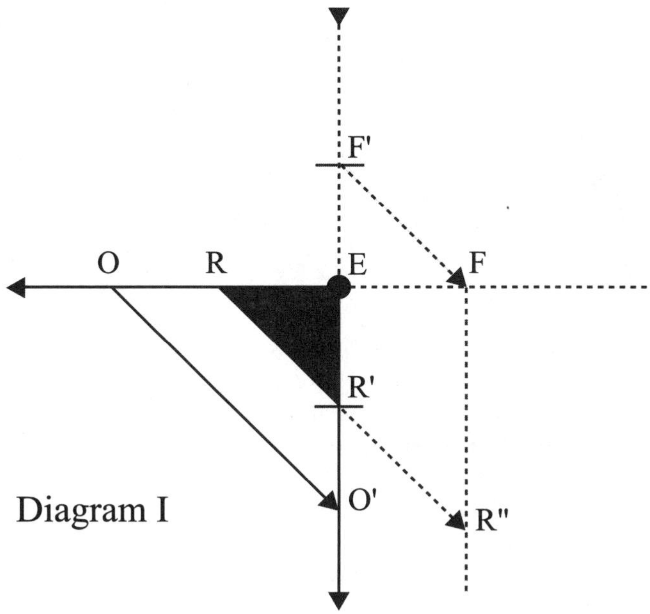

Diagram I

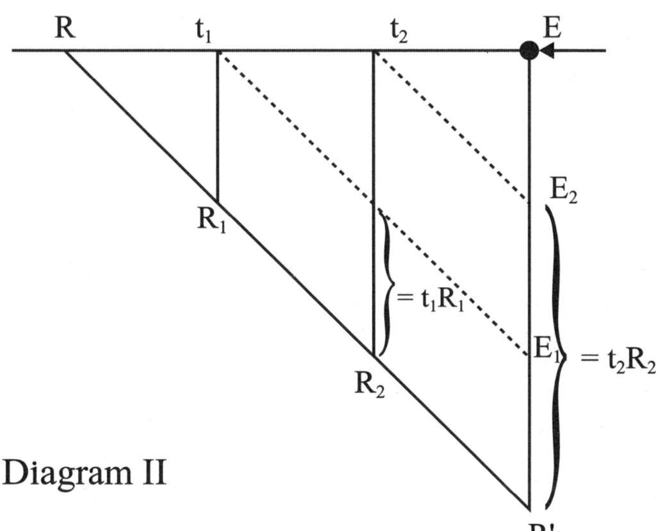

Diagram II

Diagram I

(a) *Displacement*: This diagram, which renders Husserl's preferred representational mode, is a still, representing a stretch of time seen retrospectively (or prospectively) from *one* moment or Now—as Husserl terms it, one *phase* of time. Thus whereas in the rest of this book, "phase" is used for the three aspects of time, past, present, and future, in this chapter Husserl's usage will be followed: Phases are points in a continuum (*Husserliana* X, 364; the use is reasonable since the Greek word *phasis*, connected to "phenomenon," betokens a *single* noted appearance).

The origin E is the "source point," the spring of *new* perceptions.

(b) *Horizontal axis* or abscissa: EO stands for the series of Now-points as they "run-off" to the left into the past (10). Because of the Phenomenological reduction, this axis represents an internal flux of Nows, not objective or world-time, though the two flows are analogously structured as "orthoidal," that is, as straight-line continua (appendix I). FE represents future Nows, phases yet to "run off." As E is the present Now-phase, so R represents some midpoint in the event, and O stands for its null-phase, its beginning way back.

(c) *Diagonals*: RR' and OO', the oblique lines between the perpendicular axes, represent the "sinking down" of each event-phase as a composite of the horizontal running-off into the past and the vertical drop first into primary and then into long-term memory; the arrows show the direction of projection, which is onto the vertical now-axis. Since the horizontal running-off never contains the same Now-point twice, and since the sinking away of each phase is understood to occur at the same rate, all the oblique lines are parallel. All the points in each oblique line picture the same moment in the event but at a greater remove from the current Now, E, that is to say, more deeply sunk away; hence the lines RR' and OO' are each histories of one Now as it falls into the past. (Husserl speaks of the simultaneity of all the points on each oblique line, *Husserliana* X, 230, but that is not quite accurate: each point represents the *same* impression at one time.) Similarly FF' represents a future point-event as it is projected from the present onto the Nows to come.

(d) *Vertical axis* or ordinate: The oblique lines project all the Now-points from the running-off axis onto a perpendicular line EO', which represents the continuum of phases as they are now in consciousness. In broad terms the axis represents memory below E and expectation above E. Each point on the vertical axis of Nows remembered corresponds to a unique point in the same order on the horizontal axis of Nows running off. (A scrambled order, that is, a crossing of the oblique sinking-away lines would represent a scrambling of memory, but Husserl considers that all the flows stand under inexorable laws of equable or homogeneous progress, 33.) The segment EF' represents the expectations thrown onto consciousness by the quasi-perceived Nows of the immediate future (p. 144). Though Husserl does not say this, the projection for the more distant future probably ought to go the other way, as drawn here: Expectations influence future Now-events. The point R" shows how deep a memory will have sunk down by future moment F.

It is one of Husserl's great discoveries that the memory axis is not quite homogenous but that close to the perceptual event-origin E it is distinct in constitution. The short line at R' bounds a finite small stretch of time—possibly fixed for all events and individual consciousnesses—during which memory is not reproductive but "*retentional*." Retention is a memory-mode that does not require recall or recollection but is directly attached to, and in fact *is* perception. The shaded triangle ERR' may then represent the whole perceptual-retentional episode, say a melody in its double slippage—back into the past and down into retention. The episode appears as two-dimensional because both running off and sinking down are continuous, and so the oblique lines falling away from the horizontal axis have no gaps and are more than densely, that is, continuously, packed. Thus, while the inside of the small shaded triangle represents a present perceptual event, say a melody, the unshaded part of the larger triangle stands for time filled with objects that must be recalled.

The same may be said of the futural part of the vertical axis: from E to F' expectation is not imaginative but directly perceptual or *protentional*. These two memory-modifications will be taken up in the next section.

Lest a reader think that this is dry diagrammatic stuff, let me quote a label Husserl affixes to one version of his drawing (*Husserliana* X, 365). He marks the horizontal axis, "Series of Nows (ever new life)" and the arrow at the end of an oblique line, "Sinking away into the past (pull of death)."

Diagram II

Diagram II represents only one retentional event or episode; it is therefore a blow-up of the black triangle ERR'.

If the vertical axis stands for the Now of a memorial consciousness, it catches, as was said, only one of its moments. But consciousness changes with time—a weak, provisional way of putting what will eventually become central: Consciousness *is* a temporal flux (p. 152).

The proper representation of consciousness at different times is the moving ordinate, or, drawn in one fixed diagram, a series of ordinates. It is therefore necessary to draw a second image, one in which the emphasis is on the successive displacement of Nows in successive moments of consciousness. Husserl himself drew such diagrams, which he termed "more suitable" than the better known one drawn for the lecture series (*Husserliana* X, 329–31).

The horizontal axis ER is once again the flux of phenomenal Nows running off into the past, as the arrow shows. But now we focus not only on the one phase of "primal" perception E but on three other such moments: on R (the end of the retentional tonal event), on t_1, and on t_2. RR' is the sinking away of R into retention-memory at R'; the other oblique lines trace two intermediate primal perceptual E's occurring at t_1 and t_2 and project them on the ordinate depending from the latest primal E as E_1 and E_2. (I keep saying "primal" perception, or better primal impression, because, as I have already intimated, the Now is not all that is perceived; each Now has, in addition to the primal perception, a "time-court," as Husserl once calls it [*Husserliana* X, 167], a penumbral perceptual presence.) So the point R that has no ordinate represents the disappearance of the retentional memory, while $t_1R_1 < t_2R_2 < ER'$ represents the time-consciousness for the moments t_1, t_2, E. Each Now on the horizontal line has depending from it an ordinate that represents the state of memorial consciousness for that Now; naturally these ordinates grow as the beginning of the event recedes and are greatest at the current now, at E. Of course, just as the Now-flux is a continuum, so the states of consciousness undergo continuous temporal modification; if all ordinates were drawn we would have a plenum.

Here is what Husserl regarded as the crucial element in this multitemporal picture: It is the broken oblique line projecting each earlier ordinate onto the later and longer one; the brackets indicate these projections. They represent the incorporation of each earlier stretch of time-consciousness into the present. This circumstance, interpreted by Husserl as "retention of retention," will be explicated below (p. 143). It is very important to him.—In fact it is for him the fundamental constitutional principle of time-consciousness: the intertwining of the overlapping strands of memory.

A comparison with the diagram attributed to Augustine (p. 123) will be illuminating. Husserl has introduced two highly characteristic elaborations.

(1) The horizontal line is, as I have mentioned, Creation-time, that is, real objective time, for Augustine, but it is Phenomenal, that is, internal time-consciousness for Husserl. It bears saying again that this time is to a large extent analogous to, even homomorphic with, objective time, since external time, clock-time, is constituted through it. Nevertheless the interpretational difference distinguishes the theologian from the Phenomenologist and should not be left out of account. In particular, it focuses the interpreter's mind on the problem of the doubleness of the axes. For Augustine there ought to be no difficulty. World-time runs from the Creation to the end, or from the future to the past, according to God's plan, while the human mind sits, as it were, astride of it, intersecting and taking it in at the present of perception, while carrying its remembered past and expected future along in its vertical stretch, its "distention." But what about Husserl's case, when the two axes are *both* internal and *both* represent consciousness?

At this moment we ought to recall that Husserl is at this point not intent on constructing a theory of time (however much he may *seem* to be doing just that) but on observing what appears in consciousness and how our complex internal phenomena are constituted. Two time-flows are represented because two show themselves to careful introspection. Their description will be set out in section VI (p. 148).

(2) The second significant difference is the introduction of retention, of perceptual memory—though it is not entirely certain that Augustine did not have a glimmer of such an extended presence of sensation. To be sure, he says clearly that the present "has no space." And yet, when he considers more closely how to compare the lengths of two syllables, he says: "The impression (*affectionem*) that things as they go by make in you [my mind], and that, when those have gone by, remains—that itself is what I measure as being present" (XI 27). My guess is, nonetheless, that Augustine would have acceded unwillingly to Husserl's retention, because for him the point-present is analogous to the divine "standing Now," which is out of time, but "from it, which is always present, are created and flow out all past and future" (XI 11).

In any case, the notion of retention requires further explanation.

V. Retention and Other Time-Terms

The time-diagrams are representations of the following terms, lined up here in the temporal order of their intended objects from past to future: memory → retention → primal impression → protention → expectation. Of these, retention is Husserl's central contribution to the analysis of internal temporal passage. But it turns out that retention is to be explained with reference to perception. Therefore I will begin with perception, which includes "primal impression."

Because of the mass of refinements and retractions (such as cause Husserl to say in the midst of one of his mediations on the shading-off a tone undergoes that "by no means do I want to pass off this analysis as the last," *Husserliana* X, 276), help in making sense of the texts is very welcome. The following pieces were especially useful for what follows: Bernet 1983; Brough 1975; Fink 1966; and McInerney 1991.

1. *Perception*

The first and last thing to note about perception in this context is negative. First, it is not what happens in an indivisible Now-point, for "that in Phenomenology all [reality] should be reduced to this point is nothing but fiction and leads to absurdities" (*Husserliana* X, 169). Thus perception has a certain extendedness. It will *include* retention and protention, a remarkable claim to be followed up below (p. 141).

Second, all sorts of subtle distinctions that arise in investigating perception are not of much consequence in regard to time-consciousness, in particular these three: (a) Apprehension and contents of apprehension, the cognitive schema whose dissolution was announced above (p. 130); the Now of perceiving is for many—not all—present purposes the same as the Now of a perception. Husserl echoes Berkeley: *Esse est percipi*: "To be is to be perceived." (b) The thing and its perceived qualities and properties need not be distinguished, since for Husserl "the word thing and property signify something in common." (c) Immanent (internal) and transcendent (objective) time are not to be distinguished: "Time is not doubly there" (*Husserliana* X, 272–74, 284).

Perceptual content in general, now taken positively, is what is grasped as "self-given" (ibid., 284, n. 1), which means what presents *itself* without constitutive activity on the part of our consciousness. The perceptual act in the full sense, in which thing and property mutually confirm their unity, is more than mere (Latinate) *perception*; it is (German) *Wahrnehmung*, or as Husserl writes it here: "*Wahr*"-*nehmung*, meaning (by a false etymology, p. 85) "taking-in of 'truth.'" Such complete perception is "*attention to* . . ." (ibid., 273).—Recall that *attentio* is Augustine's word for the mind's present perception (*Confessions* XI, 28). So perceptual content in general is what is given through itself in a "lived experience" (*Erlebnis*), and captures the perceptual attention. And Now and perception are inseparable: "*What is the relation of the itself-there* (*Selbst-da*) *to the* Now? What is actually now, is itself there. And what is individually itself there is actually now" (*Husserliana* X, 211).

2. *Primal Impression* (Urimpression)

Though perception will be shown to be in some sense extended, yet it has, so to speak, a center, the Now proper. This is the moment of the primal impression, the source-point (E in the diagrams) in which the generation of an enduring object begins, changes, and ends (11 and appendix I).

It is the *newness* of the primal impression that is all-important to the structure of time, for this newness is the "continuous moment of individuation" (31) that enables each Now to hold its unique temporal position as it is temporally modified and dies or sinks away (as seen in the oblique lines of the diagrams). But it is individuated temporally *even if* it happens to be qualitatively the same as other Nows, that is to say, if the sensation that fills it holds steady. Hence it is clear that what makes each primal impression new and unique is that it is an internal time-phenomenon—not really that it has external variations: Husserl says, on the one hand, that "From the point of view of Phenomenology only the Now-point is characterized as new." For

> The primal impression is what is absolutely unmodified [that is, temporally], the primal source of all further consciousness and being. Primal impression has for content what is signified by the word *now,* insofar as it is taken in the strictest sense; every new now is the content of a new primal impression. (31)

So here the primal impression defines the Now. On the other hand, Husserl indicates in a groping sort of way that the converse might hold: that the impression comes already cast in the primordial temporal form of the Now (Bernet, 103). The fact that Husserl cannot tell whether the sense-content or the time-form is prior should not be surprising after what was noted above, that even an unchanging sense-content has continually new temporal position. The circumstance points to nothing less than our temporal nature (p. 205).

Without the welling-up of primal impressions time-phenomena would be cut off at their source, which shows incidentally that Phenomenological inquiry is by no means solipsistic, totally subjective. On the other hand, the Now by itself, isolated even if continual, cannot bring about the time we are conscious of. Primal perception has what Husserl later calls a horizon, a background upon which the element becomes a whole:

> Perception itself, as the "flowing-static" present, is constituted only through the fact that the "static" Now (as deeper intentional analysis shows) has a horizon with two differently structured sides, known in intentional language as a continuum of retentions and protentions. (*The Crisis of the European Sciences*, 1937, para. 49)

3. *Retention*

Husserl sometimes calls retention "primal memory" (*Husserliana* X, 331), since it is a temporal modification of "primal impression." In *The Phenomenology of Internal Time-Consciousness* (11), where the features of retention are clearly set out, Husserl observes as follows: (a) The "bodily" tonal Now of primal impression is constantly changed into something that has been. (b) This primal "has-been," the beginning phase of memory, is retention, still a Now, "a being actually there" (*aktuelles Dasein*). (c) A ray of intention can be directed to this just-past Now, that is, the retentional phase; or it can be directed to what is in retentional consciousness, the tone. (d) Since every actual Now is subject to the law of modification, it changes, or "shades off" continuously from retention to retention. Each new retention in this continuous flux of retentions is itself a continuity insofar as it stays itself as it shades off. (e) But it also includes in itself a continuity of retentional modifications, of past actual retentional phases.—It "bears in itself the heritage of the past in the form of a series of shadings." Thus in each retentional phase the whole previous retentional "comet's tail" will turn up, to be temporally modified from its very beginning. (The two continua referred to in (d) and (e), which are incorporated in every retentional phase, will be taken up below, p. 148.)

What makes retention what it is—crucial and problematic at once—is the double ambiguity contained in its description. We might ask: (a) Is retention perception and of what sort? And (b) Is retention intentional and what does it intend?

(a) Husserl says in *Ideas* that "every *present moment* of experience has about it a fringe of experiences, which also share the primal now-form" (para. 82). In *The Idea of Phenomenology*, he speaks of "the *retention* which is necessarily bound up with perception" (68). And again, in the *Phenomenology*, where he purports to distinguish retention from perception, he says that "the constituted act, built from Now-consciousness and retentional consciousness, is *adequate perception of the temporal object*" (16). Yet he also says that in a sense the opposition between primal impression and retention is that between "now present and bygone."

The way to understand this double-speak is that retention is a trace of perception, and that Husserl's position about this problem is as Brough says, "nothing if not nuanced." Perception has a meaning restricted to primal impression and one broadened to include retention, and this latter meaning weighs heavier: Retention is impressional consciousness of the fairly immediate past. I should add that the perception of the past is a perplexity (*Aporie*) that preoccupied Husserl early on, and he made trial of a variety of solutions (*Husserliana* X, 201 ff.). The definitive resolution for him lies in this: Retention and primal Now "are together"—and, we might interpolate, are therefore both perception. But while they are together they are not themselves *now*; indeed their temporal modifications are not themselves temporal, for that would be senseless (*Husserliana* X, 333). In this last observation lies a pointer to the ultimate atemporal level of the temporal subject (p. 149).

(b) Retentional consciousness has a "peculiar" (*eigentümliche*) intentionality (12). Its peculiarity comes from the fact that it is only secondarily of a sensory object. Primarily it is of the just-past Now as now. It intends not an object but a temporal modification: that the Now is past. As Bernet says: "Retention . . . is neither an independent, intentional act, nor is it directed toward an intentional object. Retentional consciousness is an appendage of a currently present, primordial impression" (102). Therefore the Phenomenological concept of the present is hard to circumscribe: "It is neither a 'punctual' Now nor a determinately extended time-formation" (Fink, 22). It is bonded through one and the same mode of apprehension, through perception, into the before-and-after; the after-bond is retention, the before-bond protention (p. 144). The "specious present" is indeed a refractory notion!

The bonding is not through the temporal object but through time-consciousness. It is an error, Husserl says, to characterize retention as memory:

> *Memory* is an expression that always has reference only to a constituted time-object. But *retention* is an expression that can be employed to designate the intentional relation (a fundamentally different one) of a phase of consciousness to a phase of consciousness, where phases of consciousness and continuities of consciousness should not themselves again be viewed as temporal objects. (*Husserliana* X, 333)

How retention accomplishes the bonding of the Nows can been seen in diagram I: Every point of the vertical segment ER', which contains all the Now-points in primary, retentional memory, is connected by an oblique line to all the Nows of the horizontal line ER, which have "run off." Each oblique line, recall, contains the same Now as it sinks away or "shades off" into the past. Therefore each just-past Now in retention maintains continuous connections to all the previous Nows not only as they push each other back into primary memory, but also as they retain their Now-position in the horizontal flux (p. 136). Each retentional phase "encloses in itself, in a chain of mediate intentions, the entire series of elapsed retentions" (appendix IX). Each phase has its "reverberation," a "comet's tail" of phases, "though it is kept in focus as one and the same and so maintains its unity as a *this*" (*Husserliana* X, 355, also 295). Hence for each past phase (which is, recall, part of the broad present) the whole retentional continuum shifts; not only is the predecessor-phase modified, but the whole retentional series changes location and character. This is the power of "retention of retention," the (primal) memory of memory: No retention is merely replaced by a new one "lengthwise along the flux" (11), that is, along the Now-axis ER, but, once again, as each R sends its "comet's tail," its oblique line, down to project point R' onto the vertical line of phase-consciousness, the whole black triangle that symbolizes the retentional event expands, giving each fresh retention a modified, deeper history. At each moment, the series has absorbed all previous retentions and is revalued by the newest Now, somewhat like a mathematical recursion. The tonal example would be this: As I hear each new note, there is an appendage of the prior retained note that carries in it a primal memory of preceding notes, each of which is similarly linked to the ones before and modifies the melody as a whole.

This nesting of infinitely many continua within continua bothered Husserl. "Is it not an absurdity?" he asks (*Husserliana* X, 328). On consideration he thinks not, and draws just the new figure I have appropriated for diagram II. It will show that we do not have an infinite-dimensional continuum but only a two-dimensional one. The retention-consciousness continuum for each past Now that occurred at the points R, t_1, and t_2 on the Now-flux abscissa ER is represented in the respective ordinates that are fed, so to speak, by the broken oblique lines. Hence each ordinate is, as the diagram shows, included by projection in the later ones, and all are included in the present one, ER', *so that the latest ordinate contains all retentions within itself.* Thus the profusion becomes representable as a two-dimensional activity with a one-dimensional result—our retentive memory.

Retentions, which belong after all to a single sensory event, shade off eventually into emptiness. Their immediate intuitive clarity gradually ceases and the whole figure "draws itself together" analogously to spatial perspective and disappears into obscurity (9). That event occurs near and beyond the short horizontal line near R' in diagram I. The completed temporal object can now re-arise, with greater freedom of disposition and repetition, in reproductive memory (p. 147).

It is irresistible to compare Husserl's phase-terms to contemporary studies of memory, which employ the terms of information-processing. Husserlian retention is, in those terms, analogous to the "sensory register" in which tonal or visual memories are briefly held in "veridical" or unprocessed form. "Retention of retention" might be compared to "short-term memory" (STM) in which sensory memories whose pattern has been recognized are briefly held and strengthened by rehearsal and other devices. For example, a famous early experimental finding established the fact that seven is the limit of groups of *any* sort that can be held in STM. Consequently "chunking," for example, the squashing together of numerous letters into seven word-clusters, can increase capacity, since the limit is not on elements but on groups of elements (Miller, 81). Long-term memory then roughly corresponds to Husserl's reproductive memory, discussed below. Moreover, there seem to be brief neuronal system states (30 milliseconds) that are the quanta underlying primary experience, and longer neuronal mechanisms that are responsible for integrating the former into a temporal succession (Gallagher, 161); these might be interpretable as subserving respectively Husserl's primal impression and retention.

The reason that it is not inappropriate to mention the findings of a naturalistic psychology that Husserl is at pains to exclude is that he himself supposed that the objective, clocked time of internal time-phenomena might be a question for psychology. Just because Husserl had no discernible influence on these experiments, it is remarkable that his temporal modifications turn up in them—or perhaps not so remarkable after all, since the scientists too are forced to begin with the self-reports of their experimental "subjects," in fact, with introspection. Yet these later analogies should not obscure the fact that "retention" and "retention of retention" are specifically Phenomenological discoveries of Husserl. They will be the binding fibers of the rope that is the continuous internal time-flux.

4. *Protention*

Husserl speaks only briefly of this symmetrical opposite to retention on the future side of primal impressions (24, 26). Just as retention means retaining, so protention means reaching out. (I cannot refrain from collecting here once more all the words formed on the Latin verb *tendo*, "to stretch," used by Augustine and Husserl between them: distention, attention, intention, extention, retention, protention.)

Protention is a part of perception, its penumbra. It is represented by EF' in diagram I. Beyond lies what both Husserl and Augustine called "expectation," which corresponds to memory. Neither of the latter are immediately present; they require a special act to make them present, as will be explained just below (p. 145).

How is this reach into the future to be understood? One might doubt that there can be "direct intuition of what is to come" (McInerney, 116). But Husserl is thinking of a very brief duration, the momentary preview we have of phenomena about to become actual. Socrates, in establishing the cognitive modes—to speak anachronistically—appropriate to various realms of knowledge assigns what he calls "trust" (*pistis*) to the natural world (Plato, *Republic,* 511). We rely on an implicit expectation that when we put our feet on the ground, it will support us. So in the very short term we reach into the future and perceive it. Yet in a sense this perception is "empty":

> Every primordially constitutive process is animated by protentions which constitute what is coming as such in an empty way, gather it up and bring it to fulfillment. (24)

Without such trustful reaching ahead, time-consciousness would begin abruptly anew at every Now, instead of being woven into the future. If Husserl had been minded to emphasize the future—as he is not—here might be discerned the point of departure for that anticipatory "being-ahead-of-itself," that ecstasis of the future, which best defines human Being for Heidegger (p. 105). Yet there is one unbridgeable difference: Like Augustine, Husserl collects all actual consciousness into the present. Thus protention reaches into the future *in* the present. It is not truly intrinsically futural, as Heidegger means this ecstasis to be.

Retention, primal impression, and protention together constitute the perceptual, actual present. For all the acuity of observation that its Phenomenological analysis displays, the Jamesian term "specious present," which is to say "deceptively plausible present," fits it all too well. For one might ask: *how long* is the present, this Now, this phase of phases? Husserl's answer lies in the continuity of time, which will be the subject of the next section (p. 148).

5. *Memory*

As distinct from primary retention, there is a long-term, no-longer-perceptual secondary memory (*Erinnerung*), memory as we usually think of it (14). The memorial experience is complex. We remember something (1) as past, that is, not present, not now; (2) which was, however, once present, in a bygone earlier Now; (3) and which is in some sense again present in the actual Now; (4) but with a definite temporal position that marks it as past, and, in fact, as having been continually "pushed back" with each successive actual Now (27, *Husserliana* X, 178 ff.).

"Thus in memory we stand in the Now; in the object of memory we stand in the past Now" (ibid., 202). We stand in the Now because, as Augustine says, memory "is the present of things past" (XI 20). The object of memory stands in the past because it bears the mark of an actual perception that has sunk away into its definite position in the line-up of retentive-memorial consciousness (represented by all the points below R' on the vertical axis of diagram 1). The question is how this duality of consciousness is accomplished.

There are two answers, an earlier and a later one (Brough 1975, 47 ff.). Husserl evidently once held an image-theory of memory, much like the one so vividly described by Augustine. Of course, Augustine also admits imageless and atemporal memories, memories of matters of thought rather than of sense (X 11–12), just as does Husserl, citing, like Augustine, mathematics (45). But the question here concerns only memories of perceived objects. Of these the traditional image-theory, first worked out by Aristotle in *On Memory and Recollection*, says that the past perception is *re*-presented in the present by a depicting image that renders the absent object together with a consciousness of its being earlier; in contemporary terms, the image is time-tagged. Aristotle explicitly raises a question recognizable as one of intentionality: How do we see through the image to the object, so that we do not usually consider ourselves to remember the image but rather the once-perceived object? The answer is that the image is both copy *and* object; in Husserl's terms the image is an image *of*, or intends, the object (450 a 23; p. 129).

I enter Aristotle here to show how ancient and entrenched is the pictorial memory (*Bilderinnerung*) that Husserl followed at first (*Husserliana* X, 152 ff.). Then, in accordance with his tendency not to put a divide between apprehension and its contents, he worked out a new—as far as I know, a brand new—way of understanding memory and indeed imagination.

He begins by rejecting the notion that memory is pictorial consciousness. The past in memory does not involve making an image of what existed earlier (*Husserliana* X, 316, 309). His reason is that memory is a *direct* consciousness, not a mediated one. In fact there would be no way, he thinks, of getting through the opaque image to the object as past. Thus he must take on the perplexity (*Aporie*) whether memory could be simply perception preserved (ibid., 202), and he evidently comes to see a middle way: The vivid "image" of a vivid experience cannot be an analogue; the thing *itself* appears as an unmediated internal phenomenon. The representation is of identity; it is not mere pictorial similarity (*Husserliana* X, 184). This theory of nonimaginal "representation" will have a large influence on the study of the imagination, largely through Sartre's *Psychology of the Imagination* (1940); its claims will be weakened by the many picture-like features of mental imagery found experimentally in cognitive science (Brann 1991, 242 ff.).

Thus Husserl introduces a new mode of givenness for memorial objects. *Re*-presentation is replaced by "presentification" (*Vergegenwärtigung*). The word, his own coinage, means literally "the act of making present," as does its usual English rendering, which has the advantage of underscoring the missing *re*-. Although memory is still said to be reproductive recollection (*Wiedererinnerung*), in the paragraph of the *Phenomenology of Internal Time-Consciousness* where all the terms are first set (14), Brentano's theory of the reproduction of memories on the basis of images (*Phantasien*) has already been rejected. In fact, the following modes of givenness are there set out: presentation for perception, presentification for memory (and expectation), and representation for images. I might add here that Eugen Fink introduces an analogous term for retention and protention, namely *Entgegenwärtigung* (22). Its ugly literal translation would be "depresentalization." The term denotes the perceptual slipping-away to which these temporal modes are prone.

Hence we have, once more and by way of a list of the modes in which acts of consciousness intend temporal objects: presentification (memory); depresentalization (retention); presentation (primal perception); and representation for temporally unbonded images.

Thus imagination (*Phantasie*) is now excluded from the time-nexus. Objects of "mere imagination" are not before me as a genuine presence, or with their own proper Now (*The Idea of Phenomenology*, 69–70). They are temporally rootless. Experiences of the imagination are out of the context of perception and they have "*no temporal position in relation to each other*" (*Experience and Judgment*, para. 39).

Reproduction and recollection are presentifying acts of memory. Whereas the order of presenting perception is a stable one toward which we can direct our gaze, presentifying reproduction and recollection are free. We can run through memories faster, slower, in one lighting-fast stroke or in articulated steps, repeatedly or only once—without at all disordering the running-off attaching to the presentified phases. Husserl makes many subtle observations about the relation of reproduction to perception in general, and to retention, protention, and expectation in particular. But this remains the chief point:

> In contrast to . . . pictorial consciousness, reproductions have the character of self-presentification. . . . Memory is self-presentification in the sense of the past. Present memory is a phenomenon wholly analogous to perception; it has in common with the corresponding perception the appearance of an object, only that the appearance has a modified character by virtue of which the object stands forth not as present but as having been present. (28)

It stands forth in consciousness as itself, not through an image.

Memory, finally, as an act of consciousness, is intentional. Brough has worked out, in the face of the unclarity of Husserl's text, a triple intentionality: We have memory of (1) the immanent object or event as it was given us in perception, (2) the constitutive act of consciousness once experienced at the Now of that past phase of perception, and (3) the past segment of the absolute ultimate time-constituting flow, the inner time-consciousness itself. Apart from the fact that these intentionalities pretty well cover the analysis of memory given above (p. 145), the reintroduction here of internal time-consciousness itself is opportune, for it serves as a warning that so far we have been dealing with the Phenomenology of time-modifications and modes, but not yet with time and time-consciousness itself. We will be ready to do that in section VII.

6. *Expectation*

Expectation (*Erwartung*) is, like protention, briefly treated (26). It is characterized in terms of its symmetrical opposite, memory. For both of these I reproduce a "picture" of an event that I represent to myself intuitively, that is, with some sensory fullness, and to which I connect intentions either of past or of future. Therefore expectation has a temporal setting that terminates in the Now:

> Thus the expectational intuition is inverted memorial intuition, for in such a one the now-intentions don't "precede" the event but follow it. They lie as empty environmental intentions "in the opposite direction." (26)

In other words, we imagine, or better, presentify, a future event from past experience and expect its perceptual fulfillment in some subsequent Now.

Husserl adds, coming close to Augustine (XI 19), that a prophetic consciousness, one that thinks of itself as such, is in principle thinkable. It would be an expectation in which the openness that usually attends futural intentions is replaced by a definite plan so vividly presentified that it has every characteristic of being about to be, of being a future actuality. But even in such planning there is much irrelevant filler that shows that anticipation is in its nature open (26).

VI. Double Intentionality

The two-dimensional or double continuity seen in the diagram is now absorbed into the notion of a "double intentionality" (39 and appendix VIII). The horizontal axis may be interpreted simply as living, as "the stream of constituting consciousness" in which temporal phenomena make their appearance. "But living (*Leben*) is not lived experience (*Erlebnis)*" (*Husserliana* X, 301). Husserl happens here to be speaking of protention, the living experience that is a "living towards," with an anticipatory forward intention that is to be fulfilled in the future. But what he says holds of retention too: Living is represented by the abscissa; lived experience is on the phase-consciousness axis, the ordinate.

The notion of "double intentionality" is connected to the double continuity set out before (p. 141), but it is described somewhat differently and in a way more suggestive of an even deeper analysis of time-consciousness. Husserl speaks of "longitudinal" and "transverse" intentionality (*Husserliana* X, 380). The transverse intentionality, roughly identical with perception, becomes itself the temporal object intended by the enchained consciousness of retentions of retentions. These are the acts that constitute the temporal object as immanent, with *our* retentional (and later memorial) history of it. It is a history that is not one of atomic moments, because each new phase carries in it a primary memory of all the preceding primary memories *as* memories: I remember my memories of my memories of my memories. . . . The longitudinal intentionality, on the other hand, is gotten by means of the "rays" of intention that consciousness sends back up the oblique line, so to speak, as if the abscissa were retroactively projected from the ordinate. Thus is constituted a horizontal flux of "actual" Nows in which transcendent object and objective time appear and have their duration through a succession of phases. (I must enter a caveat here: Husserl, as is his way, speaks so much and so variously of this double intentionality that it is actually not quite clear to which dimension each of the intentionalities belongs. I have chosen here the interpretation most consonant with the diagrams.)

In other words, the two intentionalities are reciprocally involved. They "require one another and are interwoven with one another."

> It is the one unique flux of consciousness (possibly within a "final" consciousness) in which the immanent temporal unity of the tone constitutes itself and simultaneously the unity of the flux of consciousness itself. As shocking (if not to begin with even contrary to good sense) as it may appear that the flux of consciousness constitutes its own unity, it is yet so, and can be made intelligible in terms of our essential constitution. (ibid., 37)

And Husserl goes on to show, as I have just done above, that our glance may alight on either kind of intentional consciousness, but that they are each derivative from the other. The stage is set for a unitary view of time which, taking off from the Phenomenological distinctions detailed above, goes more deeply into the time-constituting subject itself, the ego.

VII. Absolute Time

1. *The Absolute Time-Flow*

We now come to the depths, and two perplexities in particular stand forth. A first one is that for Husserl, as for Augustine, it is a deep characteristic of time that actuality lies in the living present. *It* is new, fresh, given, intuitively fulfilled; it is, in sum, where, or rather when, we live (Derrida 1939, 137). They are not the only ones to think that. In fact, Plotinus and Heidegger are exceptional in *not* thinking it. Yet Husserl and Augustine most vividly bring the empathetic reader face to face with the dizzying possibility that we may indeed be all there, for ourselves and with the world, only moment by moment, and if we are to save our life from punctuality and recover it as a span—the way it feels before reflection—some thinking to the edge of the sayable is required. Speaking precisely, Husserl must show us not only how a Now can be so stretched as to acquire the forecourts and time-tails of lived, perceptual future and past consciousness but also how the longitudinal Now-flux is to be understood. Augustine accomplished this effect by stretching the mind "vertically" about the intersection of Creation-time and soul. Husserl, self-consigned to doing without world-time, will have to think out how consciousness constitutes time and how it renders its two strands, as in fact experienced, intelligible to itself. Husserl never tires of noting the importance and difficulty of this investigation. It may well be that in centering it on the perceptual present, he has overstepped the bounds of his own Phenomenology and made an opening "metaphysical decision" (Ricoeur 1967, p. 205).

The second perplexity is how a time-flux can be apprehended *as* temporal except by a subject that is itself in time, for here looms an infinite regress. No account of time escapes the difficulty that it takes time to apprehend time. The first straightforward solution to this difficulty is the dual notion of inner and outer time, which permits the internal sense and the external clock to check each other reciprocally. This solution is satisfactory only if there *is* external time—and in the first chapter of this book I have tried to show that there is not, at least not in the way needed. Moreover, this solution is no option for Husserl for whom the subject constitutes everything. For Augustine the temporal constitution of human beings is not ultimate, while the embodied soul is an internal world. The "inner man" can wander and search through its "spacious fields and palaces" and there find expectations of the future and memories of the past. And then he can go even further: "I am mounting up by the steps of my soul to thee who dwellest above me. I will pass even beyond this power of mine, which is called memory. I will pass beyond that I may approach thee, sweet light" (X 17). But this way of wandering through and beyond his inner temporal dwelling, the way of theology, is not open to the Phenomenologist as he descends step-by-step into the near-ontology of transcendental Phenomenology. For time-consciousness and self-consciousness are two aspects of one subjectivity, and the self-constituting consciousness is not separable into dwelling and dweller. (We see here the difference between "*being* a subject" and "*having* a soul.") So the perplexity of ultimate time will drive Husserl into trying to say what has "no name" (*Husserliana* X, 371), though it bears the adjective "absolute."

Husserl reaches the ultimate time-flow through two anterior "levels of description" that are to be discerned in the Phenomenological analysis (34–35, and appendix VI). There are three levels of time-consciousness.

(1) The experience of "objective time," both the time of nature with its physical things and of "inter-subjective," that is, common time (p. 154). This level, which includes the being-there of the objects, seems to be "empirical," meaning: not yet subjected to the Phenomenological reduction.

(2) The internal appearances in "pre-empirical time" (we are going constitutively deeper into the subject), the immanent unities constituted in their "temporal extensity," namely the retentional and protentional Now. This is the level to which the Phenomenological analysis of inner time-consciousness has been largely devoted.

(3) Subjective time "constituted in absolute, timeless consciousness, which is not an object." Here psychological Phenomenology goes over into transcendental Phenomenology, which is all but ontology; this level can scarcely be said to be merely descriptive (p. 130).

Granted the approach of psychological Phenomenology, the first two levels are relatively unproblematic. The third was felt by Husserl to be the deepest problem of Phenomenology and of the greatest difficulty (Boehm in *Husserliana* X, xxx ff.; Brough 1977, 98).

In *Ideas* Husserl says something startling:

> Time is the name for a completely *self-contained sphere of problems* and one of exceptional difficulty. It will be seen that in a certain sense our previous exposition has suppressed by silence (*verschwiegen*) a whole dimension, and necessarily had to suppress it in order to keep unconfused what immediately comes to sight in the Phenomenological approach and determines, even leaving the new dimension out of account, a closed field of investigation. The transcendental "absolute," which we have separated out through the reduction, is in truth not the ultimate; it is something that constitutes itself in a certain deep-lying and completely peculiar sense, and which has its primal source in an ultimate and true absolute. (para. 81)

Now what is startling is (1) the revelation of a *historia arcana*, a secret past, of Phenomenology, and (2) the claim that the absolute is not absolute but depends on a yet deeper absolute—which is, however, that first absolute itself. This is not a nonsensical way of breaking off the infinite regress mentioned before but candid groping for an ultimate that is not grounded in a Being *beyond* the subject but remains as consciousness *within* the subject. Husserl is trying to articulate ultimate subjectivity understood as self-constitution, of which more below.

Both Boehm and Brough connect the secret history, the discovered but suppressed self-constituting absolute, with the rejection of the "schema of content of apprehension–apprehension" referred to above (p. 130). In this cognitive schema, recall, relatively neutral givens, sensory material, or time-experiences are animated into transcendent objects by being actively taken up by consciousness. The schema is suggested by the experience of "seeing as"; for example, the same arabesque can be seen as an aesthetic squiggle or recognized as a sign, and a pink patch can be recognized as a rose or a nose. This schema is, Husserl thinks, useful to the lower levels, but at the depths it has to give way to a presentificational directness and an immanence without inserted intentional opacities such as "contents." The flux has to be, as it were, transparent to itself.

In the process of thinking out the three levels and the "absolute timeless consciousness" in which subjective time is constituted and attains givenness, Husserl reviews, as seems only right, the perceptual levels that go with each, for perception is the outer spring of *all* temporal life. He finds these four levels (appendix VI):

(1) The perception of "the whistle of a steam whistle," as it is there. This perceptual level evidently corresponds to ordinary, Phenomenologically unreduced, sensing.

(2) The perception of the content of the enduring sound itself and the tonal process apart from its integration into nature.

(3) The perception of the tonal Now and simultaneous attention to the tonal just-Now retentively united to it. Levels 2 and 3 evidently correspond to the Phenomenological analysis of internal time-consciousness proper in its two-aspects schema of content and apprehension.

(4) The perception of time-consciousness in the now-appearing of the whistle and of its extending or shading-off into the past. Here the absolute begins.

For in explicating the fourth level of perception he exclaims: "Do we not have something highly remarkable here?" What is remarkable is that I may perceive time, that is to say, just have consciousness of it, but I may also make an object of this time-consciousness, and such reflective perception will itself take time. In fact it takes just the same time as does the primary consciousness of level 2. On level 4, time is perceived at once as given and as constituted, that is, brought to intelligibility. Recall that "constitution" means exactly that: the "act-character of perception" through which the subject makes what it perceives have meaning to itself (*Ideas,* 84, n.).

Husserl had been at pains to show just this: that perception of time requires time (*Husserliana* X, 192), or succinctly put, that the perception of succession precedes constitutively the succession of perceptions. Now Husserl, in a last consideration, melds these two aspects—and gets a flux without duration. "No possible constancy can be compared to the continuous flux of the phases of appearance" that finally comes to light.

Husserl mentions several indices by which, he apparently thinks, it announces itself. A sign more on the exterior side of the subject is simultaneity, the all-embracing "all-together, the all-at-once, of the actual primal sensation" (38). This all-together, which is preserved as the impressions run off, points to, one might even say, mirrors, a "connecting form," a "stable continuity," a "singleness" (*Alleinheit*) of consciousness.

Another index is more interior. It comes out in the absence of sensory change, that "rest" which was a trouble spot in Aristotle's account of time as numbered perception (p. 46). I can, it turns out, turn my regard to the flow of tone-phases in their continuous appearance even when the sensory matter is unchanging. Even a perceptual experience that lacks the impelling life of newness, even an unchanging perceptual consciousness, has duration. Here is revealed a flow, *the* flow, without change. But what is an unchangingly filled duration? Husserl has implied that alteration is, so to speak, the empowering testimonial of duration. And so, when nothing is changing, nothing can be said to endure. But something *does* change: the phases of the flux pass off in continuous temporal alteration. Only this alteration has "the absurdity" of flowing just as it flows without change of tempo, without admitting a nonstretch, without start, change, or rest. Yet it cannot be said to have duration or persistence since it itself constitutes duration. Husserl is here laboring to describe the circularity of ultimate consciousness by his revolving meditations. Perhaps the most definitive way to put it is that of form: It is the form that stays put, the fluxless form of flow, the form comprising the primal impression, the tail of retention, and the horizon or anticipation of protention. This ultimate form is analogous to Kant's form of sensibility (p. 62), but made problematic. For Husserl comes on it not as an element of a systematic transcendental grounding of experiential science but at the unforeseen end of an introspective inquiry into the phenomena of any human temporal experience.

Husserl writes of these unthinkable necessities here, where his approach has gone beyond itself, less tentatively and more definitely than he ever does of his strictly Phenomenological observations. Truth to tell, what he is describing is the subjective analogue to Augustine's eternity. As Augustine finally leaves the dimensionality of the mind to extend himself *beyond* himself toward the Creator, so Husserl pulls the longitudinal and transverse flows back *into* himself, to find "absolute subjectivity" (36). So we may add more defining terms: The absolute flow is "self-constituting"; the constituting act and the constituted results are congruent: "The self-appearance of the flux does not require a second flux, but as phenomenon it constitutes itself in itself" (*Husserliana* X, 381). Merleau-Ponty calls it "self-anticipatory" (425).

"For all this we have no names" (*Husserliana* X, 371). But we do have descriptive recognition: This is self-levitation—from stable form to flowing consciousness, from ultimate Being to presented appearance, and in sum, from transcendental subjectivity to world-receiving time. Why is time both crux and finality? Because when we look within we find—Husserl finds—all inner appearances, but especially the centrally accessible ones, implicated in the inseparable conjunction of presence and flow, and once within, there seems for him to be no way out of absolute self-sufficiency.

2. *Newton: Absolute Time*

Augustine, as I have suggested before, does have a way out. This proto-phenomenologist finds an escape in creatureliness, in his dependence on God. So, strangely enough, does Newton, the founder of classical physics. Curious as it may seem, there is a direct line of thought forward from Augustine's divine world-time, and backward from Husserl's ultimate interior flow, to Newton's absolute, equable flux. It is irresistible to set this most exterior flow of cosmic time ever devised up against Husserl's absolute interior flow.—Perhaps it ought to be called Isaac Barrow's absolute time, since Barrow, Newton's teacher, distinguished time as an absolute "Quantum" that *precedes* motion, from time as a quantity measured *by* motion (Barrow, 204). In fact, I have come across a perfect expression of absolute time in Thomas Traherne's devotional prose-poems known as *Centuries*, which probably somewhat antedate Barrow's *Geometrical Lectures*: "For as there is an immovable space wherein all finite spaces are enclosed, . . . so there is an immovable duration, that contains and measures all moving durations. Without which first the last could not be . . ." ("The Fifth Century," 7).

Here is Newton's own, mature formulation of absolute time as set forth in the first Scholium of his *Principia* (1686):

> I. Absolute, true and mathematical time, of itself, and from its own nature, flows equably without relation to anything external, and by another name is called duration: relative, apparent, and common time, is some sensible and external (whether accurate or unequable) measure of duration by the means of motion, which is commonly used instead of true time; such as an hour, a day, a month, a year.

Absolute time logically comes before absolute space (taken up in Newton's para. 2) because it guarantees not only the simultaneity of all Nows in the universe, but also the duration of space itself. From the closing pages of the last question of Newton's *Optics* we may conjecture that absolute time might be a constituent of God's "boundless uniform Sensorium" by which he is "every where present to the Things themselves."—God's way of apprehending the world would be through absolute time, as ours is through relative time. And so even Newtonian flux would turn out to be internal—to the mind of God! But whether God's temporal apprehension of the world means that he is himself temporal—enduring rather than eternal—is a further theological question. It certainly seemed so to some of Newton's contemporaries, who found the Eternal Now absurd (Čapek 1986, 301).

What makes the unlikely comparison between Newton and Husserl somehow apt is the implied reminder that unless time is to be reduced to motion, there must be somewhere, inside or out of the human soul, a "stretching of the mind" that bears the world if not on an equable, at least on a continuous, ultimate flux. And always this last question arises: Is the subject of this absolute flow, be it God or mortal subject, itself in time or beyond time?

VIII. The Commonality of Time

The time-consciousness of Husserl's main work on time is, as its title says, *internal*. What is missing is common time, "the coordination of all particular durations into one unique 'temporal community.'" Nevertheless "the *internal consciousness of time of the primordial monad* is the origin (1) of time apprehended in the Other, and (2) of the common objective time, the time of the world" (Ricoeur 1967, 135). Husserl addresses the general question of a common world in his Fifth Cartesian Meditation on "monadological intersubjectivity." The solution is here rendered very curtly: We externalize our inner consciousness, our subjectivity, as a psyche in the world, and through the "naturalization" of the psyche we experience ourselves in a world among other such externalized psyches in a reciprocal community. Although "the Other" is constituted from the sources of my own intentionality as a phenomenon *within* myself, that is subjectively, he is so constituted as to bear in himself necessarily the same objective world (*Cartesian Meditations*, para. 56). In other words, I "equalize" the other subject to myself. One might say that, like a Leibnizian monad, I mirror the other's consciousness and the common world—and so arises one common, natural, and scientific time, the strange objective product of my subjectivity. The present of this world-time carries within it a history somewhat analogous to the reproductive memory of inner time. Here is an example. Through written geometry and physics, that is, through science, "the communalization of man" is lifted to a new level. But the written document conceals within itself the intuitive beginning from which an originator such as Galileo, a "discovering and concealing genius," had taken his departure. His theory keeps on working like a machine, but its naively deep presuppositions are closed off by "sedimentation or traditionalization." The theory-machine, with its conventional symbols, can be unreflectively operated ad infinitum, but its roots are sedimented in by the repeated writing down of a thousand refining discoveries piled on top of the original. However, the written document may be reawakened, the original self-evidence can be reactivated—and this recovery is analogous to the active reproduction of a passive memory (*The Crisis of the European Sciences and Transcendental Phenomenology*, begun in 1934 as Husserl's last, uncompleted work, para. 9 h).

It would be hard to maintain that Husserl's "equalization" establishes true intersubjectivity or sociality and a direct commonality of time between two fellow humans or among a larger group. In contemporary terms, "alterity," the direct and not merely construed recognition of an other's time, is missing. I doubt a common *time* can be anything but a contradiction in terms, though a common reference to natural *motions* is perfectly feasible (p. 12). There is, however, a very thought-provoking recent effort by Shaun Gallagher to amend Husserl's deficiency. It includes two related elements particularly pertinent to this exposition (1998, 179 ff.). The first of these is the reintroduction of content into the formal purity of Husserl's constitution of time: *What* impresses itself on us modifies our sense of time; the content qualifies the structure. The second element is the claim that such content is not always intentional: It may not belong to consciousness but may be "prenoetic"—an "excess, anterior content" not under intentional control, such as arises from our subpersonal embodied existence or our superpersonal social environment, including language and tradition. In the terms I have been using, prenoetic content returns an element of externality to Husserl's internal time-consciousness—and with it the possibility of participating directly in the time of others.

Now while it seems to me indubitable that being in the world and of it affects our sense of time—quickens and slows, unifies and bifurcates, sequences and scrambles it—the investigation of these variable experiences of time is not identical with the inquiry into our constitutive temporality. Even though preconscious and conscious contents modify our sense of time, thinking about our fundamental temporal structures seems to me to be a possibility independent of such impingements, the more surely, the more these are brought to consciousness by such studies as Gallagher's. What lays Husserl's introspection open to such sub- and superpersonal amendment is that he claimed to be giving an account of the phenomena of time-consciousness rather than to be uncovering the constitution of temporality—though that was where he ended.

∞ ∞ ∞

Husserl *ends* by re-constituting the common world-time he had switched off in the phenomenological exclusion. Augustine *begins* by accepting this time, God's time. Both are driven inward. Husserl engages in "universal self-cognition, to begin with monadic and then intermonadic." He quotes the Delphic "Know thyself" and says that it has gained a new meaning: "One must first lose the world in an *epoche* in order to regain it in universal self-reflection." He ends the original Parisian version of the *Cartesian Meditations* as he will end the worked-out German text, with Augustine's words:

> Do not wish to go outside; go back into yourself. Truth dwells in the inner man.

What both of these great writers on time find within is that the mind or consciousness is "stretched" about the perceptual Now.

Why exactly, one may finally ask, is neither Augustine nor Husserl to be placed in the company of Plotinus and Heidegger, who "temporalize" time out of a higher ground? The answer for Augustine is that the "stretching of the mind" (*distentio animi*) is not what is ultimate about the human being; we have a temporal constitution, but we can, by collecting ourselves, escape from it into God's "standing Now," into eternity. This "extention," as Augustine calls it, is unlike Plotinus's recollection into the higher hypostasis of intellectual eternity (p. 100). For Augustine never *derived* that "stretching of the mind," which by "extention" is to be collected back into God's eternity from that eternity; one must say instead that the *distentio animi* is, within the temporal Creation, an *imitation* of God's standing Now (p. 125). But this imitation is not philosophically grounded: The human mind is created, not derived.

The answer for Husserl is that even the ultimate absolute temporal flux, which transcends the temporal phenomena of psychological Phenomenology and is the deepest bearer of subjectivity, is, when all is said and done, an inference from observation, and a very nearly inarticulable *final fact*. Not only that, but if the absolute flux is itself atemporal, then the subject, the human being, is not, at bottom, temporal at all.

To wise heads this result may seem confused and paradoxical. But to me it seems that Husserl's honest and deep meditations have brought to light an uncircumventable enigma. The form or substrate of internal time is not intelligible as in itself temporal, for that leads to a confounding regression. Nor is it really intelligible as in itself atemporal, for that leads to the paradoxical notion of stasis as the form of flux. The temporality of a monadic subject that constitutes itself, a Husserlian subject, is ultimately an enigma, though a strongly delineated one. What is remarkable is that when human Being is thought of in other ways, the enigma is transformed in its terms without losing any of its mystery.

Part Two

Reflections

Part Two: Reflections

Study and thought, though not of necessity incompatible, are by no means the same. The ten extraordinarily deep, difficult, and engaging texts on time presented in the first part of this book keep the receptive student particularly mindful of this intellectual fact. I have certainly felt the danger of becoming so enthralled by their ingenuities, so enmeshed in their obscurities, and so embroiled in their implicit oppositions that I forgot to wonder what, then, time truly *is*. Now the moment has finally come to distance myself from the succession of dissonant theories and to face that question. And yet not before I engage in some additional preparatory delay.

First, it will probably help the inquiry to relate right now and as tersely as possible the gist of the theories that have been considered. Aristotle speaks of "the thing searched for of old and now and ever, and ever perplexing, 'What is Being, that is, what is Beinghood?'" (*Metaphysics* 1028 b 3). So might one, with good cause, speak of the present subject as "the thing ever searched for and ever confounding, 'What is Time, that is, what is Temporality?'" And the answer, though different when sought from different perspectives, cannot be a collage of theories, though it might well absorb not only the solutions but even more what might be called the irresolutions found in these texts—places where the authors actually express doubt or invite it by silence. Perhaps doubt is too weak a word. For each theory, it seems to me, opens up pitfalls precipitous in proportion to the profundity of its truths. Each theory has the gravity of its abysses, and it is in the discovery of these elusive depths—a discovery that follows more from immersion rather than from collision—that light of some sort begins to dawn. Therefore in the next paragraph I will add to each summary sentence what appear to me the chief lacunae in each of the theories presented in chapters two to four of the first part.

Aristotle, by understanding time as the number of motion, appears to externalize time while taking away its independent being; since, however, this number requires a counting soul, there must be some sort of—inexplicit—internal, psychic time. Kant regards time as the form of our receptivity for external stimulation; since this form reveals itself in a certain pulsing or counting, it becomes a question whether the form that is time is not itself temporal; moreover, since the recognition of the passage of time depends on spatial substance and its changes, it is unclear whence the order of succession really comes. Plotinus derives the temporal Soul from a ground or principle, the higher Intellect, from which she falls out by a kind of self-distancing in which she "temporalizes" herself; this theory requires one to accept the Neoplatonic hierarchy of hypostases, including the mode of descent from level to level. Heidegger similarly thinks of the human Being as "temporalizing" itself by self-exceeding "ecstases," though for him the ground is not trans-human but is the Temporality of the human Being itself; the difficulty is in making this standing-out of itself, this self-overleaping, nonfiguratively intelligible. Augustine is a third "diasporatic" (that is, "dispersion-minded") thinker; he accounts for our sense of time by seeing the mind as "distended" or stretched about present perception so as to accommodate simultaneously a present memory of past things and a present expectation of future things; the problem left unresolved is that of the external passage that feeds present perception and of the extension of that present itself. Husserl, a fourth diasporatic writer, solves the problem of relating the present, the moment of "primary perception," to its immediate "retentional" past and "protentional" future by giving a model for the orderly sinking away of perceptions and their intertwining with present consciousness; the deepest perplexity arises at the point where his purely descriptive approach to the consciousness of internal time reaches its limit, and he must confront the enigma of an ultimate inner temporal flow.

Second, since I will want to say that time properly so-called is the time of the "phases"—also called dimensions (Hegel) and modes (Peirce)—of past, present, and future, it seems right to give a sort of survey of them and their faculties, including only those features that are pertinent to my point (chapter one, section I). That descriptive summary will also be an opportunity for putting to use at least a small part of the enormous literature on time.

That literature tends more toward problems—trouble-shooting, puzzle-solving, and spot-checks on texts—than to the grand theories that go to the roots of human Being and Existence. It is not that these problems are less perennial, only that they are more local—though in their restricted way just as absorbing and, apparently, a good deal more provoking of argument. But perhaps it just seems that way to me, because I am persuaded that position-occupying is an approach somewhat dissonant with inquiries that concern human life. In his *Psychology* James seems to me to give a lovely and accurate simile for what I think of as thinking: "Like a bird's life, it seems to be an alternation of flights and perchings" (James, 160).

One prime example of the problems much—and rightly—agitated by scholars of time is the problem of the relation and priority of what is known in linguistic analysis as "tensed" and "tenseless" time-speech, in McTaggart's argument as the "A and B series" (p. 21), and more generally as "phase-time" and "succession-time." This fundamental distinction, which most authors accept, goes back, at least in some version, to the Neoplatonic Iamblichus (d. 325). Not so unlike McTaggart, he thinks of phase-time as flowing—he also uses, besides "flow," the Plotinian and Heideggerian word *ekstasis* (p. 103)—because events shift out of future through present into past. Succession-time, on the other hand, is a *stasis*, because events are changelessly fixed in their dated position in the succession (Sorabji, 33 ff.).

I have come to picture these two versions of time to myself as "the triptych" and "the passing point." The triptych, a triple-paneled picture, is the figure for phase-time: It has an evanescently small but vivid centerpiece, flanked by a hard-edged but fading picture on the left and a hazy but hopeful picture on the right; the three panels are hinged so as—somehow—to fold upon each other. This folding betokens the simultaneity in our consciousness of remembered past, perceived present, and expected future. The passing point is a model for succession-time: It trails behind it the actual line-segment it has generated and moves forward along a path that is predrawn but only virtually there. This diagram shows a time that has the elements of before, now, and after, but without reference to any time-consciousness; it is time in itself—nobody's time.

Since the observation that time comes in these two shapes is, whether treated as a humanly weighty matter or in the problems-and-puzzles mode, practically universal, it is necessary to come to grips with some of what is said on its account, especially if one wishes to offer a possible preference and resolution. The chief sources are two collections, edited by Robin Le Poidevin and Murray MacBeath (1993) and by L. Nathan Oaklander and Quentin Smith (1994), respectively.

A third delay, after setting out the phases and some of what is said about them, will come from reviewing them in a respect that has been more part of the implicit than the explicit agenda of this book. I am convinced that the study of time is a kind of proto-ethics, not only because reflection on the nature of time is generally good for the soul, but more precisely, because a review of the temporal sensibilities attached to the three phases, be they healthy or pathological, is bound to affect the disposition we make of the time of our lives (chapter one, section II). I would, in fact, go further:

> I believe that the reformation of our civilization must begin with a reflection on time. (Octavio Paz, "The Contraptions of Time," *In Light of India* 1995)

Finally, there is a fourth prelude to the attempt to say what time is: the collection of all that I can think and have read of that it *is not* (chapter two, section I). Then, finally, I will try to say why human time, the time of the phases, seems to me time par excellence, how the time of succession comes out of it, and so what, then, time is.

Chapter One

The Phases of Time: The Human Dimension

I. Time and Imagination: Flux and Fusion

Time and imagination, I suggested in the preface, have a certain formal similarity. It shows up in speech both about the internal images that are the most immediate product of the imagination, and about abstract Becoming, that is, the purely temporal aspect of events. From the first analysis of the nature of an image, which is to be found in Plato's *Sophist* (240), an image is understood as a fusion of Being and Nonbeing, of presence and absence: The object we imagine is positively present as a likeness and quite absent as a reality. Similarly what Hegel calls the "dimensions," that is, the phases of time, are logical Becoming, the fusion of Being and Nothing, made fluid; the dimensions are the resolution of Becoming into "Being as passing over into Nothing and Nothing as passing over into Being" (*Philosophy of Nature*, 259). In sum, images are stable pictures—Being and Nonbeing in fusion, while time is a slipping succession—Being and Nonbeing in flux.

It seems to me that the past is the exemplary phase, for it is the one in which time is transfixed for our inspection. It is, however, dependent on our capacity for image-keeping, the memory. The past is the

> time which takes away
> And taking leaves all things in their right place
> An image of forever
> One and whole.
>
> Edwin Muir, "I have been taught"

Memory-images are evidently derived from the senses, primarily from the external, distance-requiring senses of sight and hearing. But they seem to maintain a close connection to the more immediate senses like taste and touch, which act as catalysts of memory. Recall the taste of that scalloped little cake called a madeleine, which, dunked in herb-tea, lets loose floods of memory-images for the author of *Remembrance of Things Past* in the beginning of his story, and the feel of uneven paving stones that releases the successful search for lost time, the regained felicity of past moments, in which Proust's book culminates and where, with a wonderful circularity, the author's writing begins.

Imagination-images, fantasies, are also derived from the senses, if not in token then at least in type. In other words, though they may image particular entities not ever actually sensed, they have our human sensory mode. Therefore—it is no news—memory and imagination belong together, though it is not easy to say which embraces which (p. 214). So it is no wonder that the past shares not only in what is positive about images, their likeness to an original, but also in their negativity, the absence of that original. That negativity has in fact three facets: What is past is ever more bygone and gone, done and done with, unreal; what is past passes ever more into a nonveridical imagination-image, a semifiction; what is past may go into total oblivion, into traceless blank absence.

The future is similarly affected with negativity. Although it is the phase of purposed action, what is future is hazily indefinite unless shaped into figures by the fictive imagination. And even the present, the time of vivid novelty, denies itself to appropriating reflection, because "thinking about" is a two-step turn, and by the time we have bent our thought back on a Now, it no longer exists.

The point here, that time and imagination are related through the way Being is implicated with Nonbeing in Becoming and in images, requires clarification, which will be attempted later (p. 191). But meanwhile, I might point out that it is only in articulated reflection, in analysis expressed in language, that the negative aspect of images and of Becoming is explicit. In itself, unrecognized and unanalyzed, an image is all-positive, plumped into psychic or natural space, all there, without the slightest mark of deficiency. In itself, time is curtailed to mere punctual presence, without bygones or yet-to-be's. It is the time we imagine the animals to live in:

> But these have never trod
> Twice the familiar track,
> Never never turned back
> Into the memoried day.
> All is new and near
> In the unchanging Here. . . .
> Edwin Muir, "The Animals"

Both the shapes that lie in space and the animals that dwell in the present may *be* the past, in the sense of being the product of a past process or the front of an evolutionary passage, but they *have* no past, since they do not utter—and so, we may infer, do not think—Nonbeing and negation. Thus it makes sense that animals cannot be trained not to do something except by immediately punishing them for something they have already done (Deacon, 86). Children, too, cannot obey prohibitions before the word "no" comes to them, though when it comes, it comes with a vengeance. The intense and strenuous use these two-year-old terrors of negativity then make of a word in no way easier to say than "yes," which little affirmative is nevertheless not in their vocabulary, seems so much more to be their own contribution to their incipient personhood than the natural unfolding of their little positive natures; "no" seems to mean something more even than a declaration of independence: an aboriginal delight in negativity. This is, moreover, also the stage when children delight in recognizing images, when they themselves are all but little images, that is to say, accomplished and constant mimics, and when memory becomes expansive and expectation demandingly specific. These are the miraculous first displays of our human triple capacity: for recognizing that something absent can be depicted as present, that something bygone can be now recalled, and that what is not can be meant and uttered as well as or even better than what is; in sum, we are able to have and interpret images, to live consciously in the phases of time, and to think and speak negatively. My guess is that these capacities are really triune, three-in-one. They may be the root of our humanity—and perhaps the subject of another book. Here only two of the three capacities, time and imagination, are to be considered, but it was necessary to point out that there is a missing third.

There is another large consideration forgone here: the world. For present purposes I simply accept that there *is* a world; I mean that we live in and are confronted by stubborn but also changeable givens, coming to us adventitiously—"without our consent being requisite," in Descartes's phrase. This world is external, that is to say, it stretches away from us, and it is extended in itself, which means it is spatial. It displays variability that seems to happen latitudinously all over space but is received by us longitudinously in time. The face and front that the current world presents to us is interpreted by *us* as occulted past, past entombed in the current spatial spread. But whether it has a past in and for itself must remain a matter of conjecture. My guess is that the world is temporal only by our grace. But that view must be admitted to embrace an enigma: How is non-temporal variability to be understood? As always, the question must be which enigma we can best live with.

In the first chapter I will, then, try to pick out the salient features of the phases (I) and the dis-eases to which they are prone (II). The order in which the phases are given is not without significance. Bacon says in the Preface to *The Great Instauration* (1620) that "Time is like a river, which has brought down to us things light and puffed up, while those which are weighty and solid have sunk." Here the passage of time is from past to future, the Now is swept forward. When the future is seen as an oncoming doom or blessing, the flow of time is *from* the future; the Now is pushed backward: "Long countless time brings forth all things obscure and hides them once they've come to light" (Sophocles, *Ajax,* 646). Some people apparently think that only the present is alive and has priority of place, while others think that it is not time at all; so Spinoza, in his *Hebrew Grammar* (ch. XIII), says that the Hebrews have no present tense, because they regard the present as merely the end of the past and the beginning of the future.

Here, true to my sense that the past is most truly time, I put it first, then go to the least determinate phase, the future, and end with the phase that is most problematic in that it is somehow not "a" time at all, the present.

1. Past and Memory

What, then, is the past? "Past" means "passed on," bygone. Events are over, things are gone; they are past or in the past. Should we mean by past the time that has gone by or should we mean the events or things that are over and done with, as Greek and Latin usage implies—for example, Augustine's normal word for that past is "things gone by" (*praeterita*, p. 120)?

A conundrum familiar to those who puzzle about perception is whether the tree falling in the uninhabited forest makes a sound. It is paradoxical to imagine but it is possible to think that it might. So too with time:

> This is the first thing
> I have understood:
> Time is the echo of an axe
> Within a wood.
>
> Philip Larkin

It is possible to think that time is a tell-tale trail events leave in the world, a sensory, though perhaps unsensed, ghost of the deed done. But it seems to me that such echoes, such events to whose nature it belongs to be past—that is, present as echoes but past as the sound of an axe—can occur in the world only for beings that recognize images *as* images, as present absences. It seems to me that pastness is not another attribute such as can show up and be pointed to in the real world. The question might remind one of an old and ever-unsettled philosophical quarrel about existence—whether it is a "real predicate" of things, whether it modifies their character. But the problem is actually quite different: Pastness might conceivably be a predicate of things in memory that are only half-real, but it cannot possibly belong to things in the world that are real. Hence, pastness is not a predicate of real things in the world but a mode of being and nonbeing in consciousness. One might ask: What about events and things of which we have never learned or which we have hopelessly forgotten? Well, we have a sort of universal coverage for the world that insures that all the imaginable but in fact unimagined happenings are in a courtesy-consciousness, as it were; we engage in an encompassing positing of all the world's alterations together that registers them *en masse* and sight unseen. And that is how we impute to the world a past beyond our own cognizance. Aside from the world-embracing but blind act of the supposing imagination, we really have no clue how our nonconscious surroundings manage their variations.

(a) But *for us* there are surely things in the past. Then where are they? Where is the "scroll of elapsed presents," as Mead calls the past (Mead, 30)? It is surely not the case that they are simply gone, as melancholy is apt to lament: Where is the all-wise baby that has disappeared into the bright but this-worldly little boy? Where is the golden age that has settled into a mundane present? Melancholy says "gone," but memory says "ever-present."

Perhaps, then, as was intimated, the past *is* here and now, really in the world, though only implicitly, as the facade of a process, a facade that bears on it the marks and indicia of a trail of time leading up to it. If the present front was reached by temporal inertia, that is, by mere duration, the facade might look like those worn hieratic piles, the pyramids, that Hegel so despises for *merely* enduring, but if the now merely implicit process was one of progress and supersession, almost all the visible front of time would be recent and evidence of bygones would be dispersed, buried, or irrecoverably pulverized. This temporal front conceals the past on its immediate face but reveals it to investigation, particularly to the digging disciplines of the old, to paleontology, "the account of the long-ago," that is, of old forms of life, and to archaeology, "the account of the ancient," that is, of old forms of civilization. Bringing the past to light by digging deep is a satisfyingly spatialized view of time: "Deep is the well of the past," says the first sentence of Mann's *Joseph* novels. The strangeness of this "discovery" of the past was vividly brought home to me when, in my former life as archaeologist, I was sitting in a storeroom in Athens, piecing together a pot that was, we reckoned, 2,700 years old, and the ever-attentive director of the excavations stuck his head in my door and demanded to know "What's new?" I laughed and showed him two sherds.

A term for the depository of the past that I have just described is "external memory" (Bloch, 215 ff.). The identified visible remains from sherds to monuments, together with surviving tales, accounts, and documents, and the histories later based on them—all these constitute external memory. It goes without saying that such memory is not always factual, that is to say, an evidence-guided recovery of bygones: King Alfred the Great, who was documentably very real, is to most of us a dim and schematic figure, while King Arthur of Camelot, who is probably unreal, is a vivid and poignant person.

As one would expect, the discovery of the past has itself a history (Schnapp, passim). From treasure trove to stratigraphy and stylistic analysis to absolute dating by physical means, the orderly uncovering of the "shape of time" is itself developing, which reinforces the probably not-much-disputed notion that the public past is a human work. If archaeology is grass-roots history, then the evolution of shape and the articulation of generally operative reasons for stylistic changes, such as the "esthetic fatigue" of customers or the hard-to-suppress tendency toward deforming extremism of artisans, is surely of primary interest in giving a connected picture of the past (Kubler, 62 ff.). Some of the most copiously preserved artifacts, such as ancient pots and modern coke bottles, are so suggestively anthropomorphic in what archaeologists innocently describe as their mouths, lips, necks, shoulders, bellies and feet, that they plausibly mirror tastes in the human, particularly the female, figure—witness the Mae West-shaped early classic coke bottle and its less curved successors (Gilborn, 13 ff.). The point is that whether the rule of development is generally aesthetic or concretely human, common artifacts show evolutionary lines, in which time seems to take on shape, and I cannot think of a study better designed than pot-archaeology to raise reflection on an issue omitted in this book: how human beings, each of whom is the individual originator of time, come to position themselves in a common time.

The reasons people have for more or less meticulous recovery of external memory are many: the romance of discovering a lost temporal country (Lowenthal, passim); the need for heroes—recall that the Greek muse of history is called *Clio*, "Fame"; the imperative for a people to remember its bond with its God—witness the biblical *zakhor!*, "remember!" (Yerushalmi, 5); the desire to preserve a collective memory (Halbwachs, passim).

So also is there a multitude of "concepts of time" that seem to govern the organization of public memory: the Eternal Return, the indefinite repetition of archetypes that might be called the reflexivity of time, the device by which societies "refuse history" (Eliade, 91); the *kairos*, the singular irruption into human time of the divine, which gives history both its dating epoch and its meaning; the episodic account of events that does not locate them in series through time but in parallel over space, such as is employed by the first historian, Herodotus (Wilcox, 51 ff.); the fairly recent triumph of an absolute, linear, universal, and secular chronological framework for all incidents whatsoever that are joined into "historical events" within it (Kosselleck, 106; Schneider, 200); and finally the attempt by Heidegger, not altogether convincing, to overcome the divide between the material remains and the human past in the notion of "repetition," *Wiederholung* (*Being and Time*, para. 76; Ricoeur 1988, III:78).

But all these external memory modes are grounded in personal memory: "The remembered past is both personal and collective. But as a form of awareness, memory is wholly and intensely personal . . ." (Lowenthal, 194). Even "collective memory," our common social memory, requires individual recollections (Halbwachs, 38). Thus without personal awareness and recollective effort, the memories implicit in artifacts, traditions, and myths, and made explicit by public remembrancers such as historians, are unremembered memories, no memories at all.

(b) So having said where the past, all that has gone by, is not, it remains to say where it is—and that is in memory. Probably that is just what people already know, though it isn't quite what they usually say. For they often talk as if the past were in the world, as if being chronologically old were the same as being of the past. But if memory is where the past is, then a few words are in order here about our faculty for having a past and for giving the world a past, especially about its more problematic phenomena.

First, the past object, the remembered object, event, or thing, is rightly thought of as a *factum*, a done thing. Yet memory is notoriously unfactual and often counterfactual. Even "flashbulb memories," which seem so unquestionably imprinted, are often false (Schacter, 201). So our sense of pastness and our sense of past things is divided: As bygone, we think of the past as fixed and determinate; as remembered it is labile and indefinite. This indefiniteness, as a weakness of our memorial capability, is quite separable from a certain ultimate inherent indeterminacy that might be ascribed to the events themselves. Hannah Arendt tells the story of Georges Clemenceau, who, engaged in a friendly discussion with a representative of the Weimar Republic about what historians might write concerning German guilt for the First World War, said, "This I don't know. But I know for certain that they will not say Belgium invaded Germany" (Arendt, 239). It is not large, brute facts, but motives, meanings, and micro-happenings that may be in themselves indeterminate. A telling example with respect to motive, true to life, appears in G. K. Chesterton's detective story about anarchists, "The Man Who Was Thursday": "We are not murderers, we are executioners," one of them, Thursday in fact, says at a meeting. So too in Toni Morrison's *Song of Solomon* a member of a terrorist group that happens to call itself "The Days," the one called Sunday, gives a similar interpretation: What they do is not murder but a balancing of evil. Some aspects of the human past may indeed be descriptively undeterminable as facts, and so be *in effect* indeterminate.

Is memory *ever* self-verifying, veridical? Hume thinks so, because he relies on the superior force and vivacity that distinguishes memory from mere imagining (*Treatise on Human Nature*, 1739, I, iii, 5). Although it is manifestly false that the representations of memory are always more vivid or even more sharp-edged than those of the imagination, in general the degree of certainty, of unmediated self-evidence that does seem to inhere in much memory, is taken, however unfoundedly, as an indication of credibility. Moreover, since sometimes our memory is externally verified, it is for practical purposes, but even more for the recovery of our inner life, not irrational to trust it (Smith 1966, 28 ff.). Nonetheless, all in all and mainly, the memorial past, the past proper, is *en gros* determinate and in principle knowable, but in human fact, not so. Indeed, the question whether in cases of early trauma memory is truthful, whether recovered memories are ipso facto veridical or are modified or even invented under the influence of advocacy, has recently become acute as a legal issue, because what is at stake is the satisfaction of the alleged victim and the criminal liability of the alleged perpetrator of the deed that caused the trauma, usually child abuse (Brenneis, passim). The past, robust though it be as an inner possession, is very delicate as a public product.

Second, the past is thought to be irrevocable. It is exactly the past that is safe from "the depredations of time," and it is the past that cannot be undone. Its irrevocability is, of course, related to its determinacy. It is only because we suppose that, however little of it we may have in memory, the past is in fact entirely fixed, that we think of it as not symmetrical with the open future, as not open to amendment, not even by God.—In Judaism this respect for things once done goes so far as to proscribe a prayer that something should not have happened, since it mocks God by asking for a logical impossibility (Dummett, 117 ff.).

And yet—. F. Scott Fitzgerald's "elegant roughneck," the great Jay Gatsby, when told that "you can't repeat the past" cries out, in all the vitality of his innocence: "Can't repeat the past?. . . Why of course you can!" (ch. VI). His claim is that the past can be revoked and undone as well as invoked and redone. Nothing in his story proves him in principle wrong, the concluding chapter intimates; his faith is the moving principle of every renascence. And of course, much memory works in just that way: It recollects an amended past, and we proceed to live by it.

Third, the past is present, that is to say it is *now*, for when else could it *be*? The perceptions of every present become time-absorbed and time-tagged. The "scroll of our elapsed presents" becomes thicker and thicker, but we do not need—and here the scrolling metaphor fails—to unscroll it successively to reach episodes even far gone into the past. Real objects go out of sight and some become internal images, and the images as they overlie each other make up the scroll that represents continuous stretches of time mostly out of mind, but its pictures are directly accessible, here and now. So the past is in memory in two ways, as accessible and as accessed, potentially and actually, as something we presently possess though not necessarily *as* present. But this dual mode is just what defines a storage facility: a place where something is, one trusts, in one's possession at any moment and from which one may retrieve it. Whether this storage capacity called memory maintains a complete record of all our perceptions and experiences is not known. In the 1950s Wilder Penfield thought that by electric stimulation of the brain he could elicit a near complete and accurate record of long-forgotten events, but the veridicality of the memories so stimulated proved to be doubtful (Schacter, 77). Yet there is little doubt that memory storage is more often orderly than scrambled; stretch by stretch of duration, the order of events and the appearance of things on the scene seems normally to be coherent, even when there are lacunae.

So from Augustine on, talk of memory as a treasure-house or, more prosaically, as a storage space, has been regarded as more than a metaphor, because it is descriptively so apt. There is a huge literature on the faculty of memory; it is, to be sure, not much concerned with the question of time, which often does not even appear in the index. The reason is that time is not a problem in naturalistic memory studies but is rather the starting point. The ontology of time is bypassed by identifying it with process; thus word and problem disappear together (Michon, 40). Nevertheless, memory studies help in thinking about the place of the past.

A guiding distinction in such studies is among access to memory, the memory itself and the memories that it contains—the key, the container, and the contents. There are multitudinous roads of remembrance, ways into memory: spontaneous remembrance, directed recollecting, musing reminiscence, reminding memoranda, mindful recognition (Casey, 88 ff.).

The mechanism of memory itself is conceived of in terms of various models. It may be like a date-stamping machine that time-tags each perceptual event, or like a filing cabinet that is already predated and organized sequentially, or like a fading photograph that indicates time by waning vividness, or like an archaeological dig where dating inferences are made from context, or like a book with cross-references from which one can reconstruct the order of publication, that is, of perceptual occurrence (Friedman 1990, 28). Sometimes neurophysiologically based systems are distinguished: "procedural memory" for motor skills, "semantic memory" for languages and facts, "working memory" for temporal order, "episodic memory" for personal experience (Block, 72). Often the memory is analyzed in terms of its structures and the depth as well as the capacity of their levels: immediate and fast-decaying sensory or "iconic" memory, short-term memory where all that is needed for present working purposes is stored, and long-term memory, our deep storage (Klatzky, passim). These models are, of course and of necessity, figurative. Metaphors structure much of our experience in any case (Johnson, 65 ff.). But when we represent interiority we have practically no means except metaphor, since in externalizing it, we are perforce spatializing it—witness the key diagram of this book (p. 123): The soul in utterance assumes a body.

Every naturalistic investigation into memory takes its initial categories from introspective reports, and even the most ingenious experiments usually rely for testing their hypothesis, be they models of procedures, systems of levels, or brain structures "subserving" the phenomena of memory, on subjects obeying instructions to use their memory and to report results. Consequently the functional aspects of memory so hastily surveyed above—types of mental imagery, capacity, accessing, processing, and disorders, are well and recognizably discerned. But somehow the main question for my purposes is begged: how the mental images that preserve past sensation come to be "time-tagged," what it means that we carry durations in our mind so that what is no longer is not only present but often carries with it a temporal location that places it in a sequence of memories and a time-distance that attaches to it a duration of pastness that may be estimated. The sense of duration is indeed the subject of many investigations, but they *start* with the notion of temporal information and really only supply theories of its gear, so to speak, the hypothetical motions that might explain it (Frazer 1989, passim).

There is the aforementioned famous "Excursus on the Sense of Time" in Mann's *Magic Mountain* (ch. IV) that at least describes the phenomenon of duration in a vivid way: It is not always true, as people commonly say, that full time flies. Emptiness and monotony may stretch the hour and make it "boring" (*langweilig*, literally, "long-whilst" in German); but memory makes the years of empty time seem brief, even evanescent. Conversely, when time is full of incident it flies by the hour, but its long-term passage seems stretched out and densely laden. Mann's observation suggests that it is the thickness of the memory file that produces length of duration and depth of location. The memory is to be imagined as a sort of image-laminate.

Perhaps I should, finally, mention a notion of the past that is beyond memory as described here, that is, conscious memory. This is the notion of a "prenoetic past," which comprises all that we are not mindful of but that conditions our conscious self: congenital knowledge of our distinctness from others as well as the conditioning received from our environing sociality, the self-perception of our bodies as well as unconsciously exercised learned skills—in short, whatever has "operant existence in the interrelations of body and environment" and what is "*always already*" there for us (Gallagher, 125, 191, p. 155). We might indeed call all this our past, but we might also call it our present condition as seen from the perspective of human development and its preconditions.

2. *Future and Expectation*

Of the three phases, the future is the one that *is* in the least sense. What is to come does not only not exist, as the past does not exist, but it is not even a determinate nonexistence, as the past *is*. It does not seem to be quite right that, as Maxine Kumin puts it, "The time on either side of Now stands fast." On one side of the Now, there is nothing that can be tethered down.

Consequently, there is usually no special faculty assigned to this realm of indeterminacy. For what Augustine calls "expectation" is not a discernibly independent faculty like memory and its store of time-tagged images. We seem to reach the future either by prudent conjectures that depend on art and experience, or through the images of the imagination, sometimes shaped in a purely contemplative mood, but more usually by fear or desire, or through the intellect by propositions that are logically very problematic.

Thus Bertrand de Jouvenel, the founder of futurology, the art of large-scale conjectures about the future, observes that "There is no desire without object, and the object of a volition is that a fiction of the mind become a fact." Now images referring to the present or the past are called "representations" because they represent *facta*, but a selection of certain other, more mysterious fabrications of the mind, fantasies or *ficta,* serve as the source of systematic action: This image summons a future reality: *fictio quaerens veritatem*—"a fiction seeking truth" (Jouvenel, 25).

Thus the imagination that is furnished by bygone presents in turn furnishes feared or hoped-for futures, acting as a kind of projective memory. That projective memory is sometimes but not always "futuristic," that is, informed by stylistic hyperboles taken from current technology. It is sometimes backward-looking and revivalist. Readers of utopian literature will recognize the types: the futuristic technological dystopia (for example, Zamyatin's *We*) and the retrograde pastoral utopia (for example, William Morris's *News from Nowhere*). But in whatever realm expectation finds its material, the memorial imagination guided by deliberate volition gives it its shapes. And these shapes can never, by the fact of their shapeliness, be futural; only the time assigned to their realization is in the future.

There are, of course, those who believe or claim that the supposed indeterminacy of the future is only a function of human ignorance, and that all events are in fact fixed either by God's foreknowledge or by the laws of nature, human and nonhuman, causative or formal. The neat element of this view is that it makes the past and the future symmetrical about the present. The loss is that the present, which under the aspect of the open future is the momentous phase of human decision, becomes a meaningless and impotent ghost of a divide gliding through an undifferentiated duration impervious to human intervention. Thus past, present, and future are in human effect abolished. Now it does seem to be the case that unpredictability, that is, the human inability to foretell the future, is no proof of its indeterminacy.—This observation holds no less for theological predestination, in which the souls themselves do not know their predetermined fate, than for chaos theory, which deals with phenomena that, though strictly determinate in having a future specifiable by mathematical equations, are yet quasi-inherently unpredictable on account of something more hopelessly incurable than human ignorance, namely human finitude (Kellert, 49 ff.). Yet for all that, it belongs to human time-consciousness to interpret the unpredictability of *this* world, at least, as betokening openness for the future and freedom for us; thus hope derives its energy. And this view again confirms the Augustinian simultaneity of phases: concern with the indeterminate future is *present* concern, whether it determines the initiation of action or remains mere contemplative dreaming.

The future, or rather the expectation of it, seems to come in five forms: prophecy (sacred and secular), science, predictive speech, and "anticipation."

I omit prophecy, knowledge of the plan of Creation, without prejudice as to its possibility. Science, the rational ability to predict natural motions and processes is, of course, also beyond this book, though I want to recall here, because it has been alluded to before, the determinism of the side of the "Gods" in the battle about the role of time in Relativity (p. 11). This is not the causal determinism in which the future, though fixed, is yet to come in time. It is instead the wonderful notion that the world is all before us—and behind us, laid out in space, while we move across it along our world-lines as temporal sightseers of the absolute, or more modestly, "like separate flies crawling over a tapestry" (Sir James Jeans, quoted by Frank, 392).

Futurology, a secular prophecy, began, as I said, with Jouvenel's book *The Art of Conjecture*. He made it quite clear that certainties—personal expectations, public structures, contractual assurances—were the setting in which future-telling, that discerning of "strong and powerful tendencies," can take place (114). Four stable systems that make predictions possible have been playfully dubbed: (1) Planets—because these are mechanistically determinate; (2) Plants—because these are developmentally regular; (3) Plays—because they are rule-governed plots with room for human ingenuity; and (4) Plagues—because they are probabilistic catastrophes (Boulding and Boulding, 40).

The interpolation from discerned trends became the basis of futurology after the mid-century, and the most curious melange of true and false prognostications resulted from the application of "the law of more" to then-current events: the prediction of the decline of all-out war on the one hand, and on the other the now grimly hilarious notion that *our* pressing problem might be the virtuous disposal of large amounts of leisure time. Information technology was vastly underpredicted (Kahn and Wiener, 368, 198, 93).

Futurology might be called public or external expectation, in analogy to external memory. Its success depends, to be sure, on the aforementioned stability of major frameworks and the macrotrends arising within them, but these in turn depend on the small-scale tendencies of individuals. What would happen if the human soul became statistically unpredictable, failed to desire mostly along inertial lines and to imagine mostly what already exists—of that there is no artful conjecture to tell. But it seems to me that such an event is at every moment a real possibility, and books like this are written to make it actual.

A fourth mode of future-telling is simply to make futural statements. Such ordinary, amateur speech predicts, though not in the artful sense of assuming that the future can be spoken of and that there is a profession that knows how. The problems arising about futural statements, apart from their predictive value, are much deeper than those of futurology, for they are problems of logic. It has to be mentioned that many languages have no future tense, and also that in the English auxiliary forms "will" and "shall," which serve us in forming the future, the original element of intention and obligation is still often evident: "I'll do it tomorrow" (Jesperson, 50). But there are plenty of ways that even in English, as easily as in Greek, mere unwilled futurity can be uttered. It was Aristotle who first inquired into the status of such future propositions (*On Interpretation,* ch. IX, p. 53). His question is: Are statements about the future *now* true? He argues along these lines: Both because there are accidents of nature and because deliberation affects the course of human affairs, the future is not determined and not knowable. Therefore, while the logical law of excluded middle, which says that propositions are either true or not true and there is no middle possibility, holds for propositions about the realm of fact, that is, the past and the present, it does not hold for statements purporting to be about future singulars, that is, future facts. For if we posited that a true (or false) statement *can* be made about the future, that would entail that the future *has* to be in accordance with one or the other. For a true (or false) statement is one in accordance with the way things are (or are not). But that would mean that there was neither chance nor any point in deliberation, for the future would have to be thought of as predetermined. So propositions about the future cannot be true or false statements. Nothing could more elegantly show the seductive nullity of this phase than our inability, try as we will, to speak about it declaratively.

The fifth and last way of going into the future is, finally, ontological, that is to say, concerned with saying what the future as future *is*. Heidegger is the one philosopher who tries to give a deep account of the power of this otherwise fuzzy phase (p. 105). The future is not for him a Now that is to come, nor the as-yet-missing though faintly pre-scribed part of a linear progression, nor one of three longitudinally strung-out phases. It is rather a way of being authentically, called "anticipatory resoluteness," *vorlaufende Entschlossenheit* (65). The German words signify that human Being is properly such as to run ahead of itself in a mode of existence that is both resolute and open, literally "unlocked." It is characteristic of this open resoluteness that it is receptive of and stands up to its most distinctive possibility—death. For "Only in dying can I to some extent say absolutely 'I am'" (Heidegger 1979, 318). The resolute human Being runs ahead of itself and in doing so lets the future come toward itself, not awaiting it passively but living toward it. The future is thus not a phase of time but a feature of existence. To find this analysis not only deep but true as well one would have to accept (1) that all normal temporal experience is indeed an improper, inauthentic way of living with time, unrectifiable by any ordinary wisdom, (2) that "forerunning" and "letting come toward" are not spatial metaphors but literal descriptions of modes deeply distinct from ordinary purpose and receptivity, and (3) that death, the terminus marking human finitude, can and should indeed be looked to as a consummating possibility rather than feared or welcomed as the cessation of earthly opportunity.

3. *Present and Perception*

Almost everyone who thinks about human time, thinks of the present as the primary phase. That is true even of Sartre, who borrows Heidegger's "ecstatic" Temporality, but puts the accent on the present ecstasis, "not because it is ontologically prior, but because it is in the present that I am in the presence of a being that is not myself" (Sartre 1943, 176 ff.).

Figuratively speaking, the present can be seen as the cusp of time. A cusp is a geometric singularity formed by the crossing of two curves, such as is seen in an ogival arch, whose sharp tip is a point that has a double allegiance, to the curve segment coming and the curve segment going. So too the present, taken merely as one phase, is imageable as a point that is the last of the past and the first of the future, and is yet elevated above them, outstanding in its crucial contribution to human life.

For, functionally speaking, the present is the moment of fully activated existence, the intersection of consciousness and world. The present is when perception takes place, and the strangeness of this, our condition, is almost too strange for expression. *Now*, whatever that quantum of time may be, is when this world just as it is now and will probably—with the rarest exceptions—never be again, is before me, present to me as I am now and will probably never be again. We live on the punctual cusp of time. If the face of a watch represents the durable motion of the world, this motion is there for us at only one evanescently small position of the fastest-moving hand; the rest is in our past or future.—Almost the whole dial shows when the world is absent for us.

It may be a consolation, though one that compounds the wonder and really contains a conundrum, that these Nows of presence follow each other densely, even continuously, and so make up the duration that gives extension to our life. For we ordinarily live through stretches of Nows so melded and amalgamated that it is possible only in retrospect and by abstracting thought to cut the duration and isolate any particular Now (p. 209).

Three more specific questions seem to follow from this—deliberately—estranged view of the present. One is what in human and worldly Being might account for the aboriginal coupling of Here-and-Now, a coupling that characterizes the present and happens also to define existence—for in one chief usage "existing" means "being here and now" (p. 203). Why and how are Here and Now combined in our existence, though they are so very different, above all in this, that our Now is unique while our Here can recur? For, though the past can probably be brought back into our existence in *some* signification of the term "again," yet when closely regarded, no Now is exactly like any other, if only because the second Now would have a memory behind it marking it as a reiteration. Places, on the other hand, at least mere coordinate location, can be revisited (Gale, 74 ff.):

> In space things touch, in time things part. (E. M. Forster, *A Passage to India*, ch. XXII)

Yet though—or perhaps because—the Now is very narrow and in itself quite colorless, while space is very wide and very various, they seem to mate: Moments are differentiated by places, and spaces are made memorable by moments. The easily corroborable observation that all those old enough to have experienced the assassination of President Kennedy know exactly *where* they were *when* they first heard of it is prime evidence.

The second question is whether, having related the present firmly to perception as its faculty, we may speak of the present when perception ceases and nothing is present *to* us. And the third question that is ever present when the present is thought about, is what its extent might be. If it is *a* time, a stretch of time, our abstracted model of it as a partless point lying on a linear time-continuum is violated (p. 40). Yet that an evanescent quantity like a point should meet the fullness of the world in perception is unintelligible as well. So even before we meet the concrete problem of how long the specious present must be to bring about consciousness of an object of sense, we are confronted with the question why the mathematical Now doesn't jibe with the human present.

These deeper questions will be taken up below and again at the end of the book (p. 203). A review of the mysteries of the present phase does, however, immediately lead to some clarifying distinctions. For this central phase appears to be thought and spoken of in at least five different ways: as (a) Now, (b) instant, (c) present, (d) moment, and (e) *kairos* and *Augenblick.*

(a) "Now," the adverb that in a book on time cannot help but turn into a noun (and is therefore capitalized), may ultimately be etymologically akin to "new." It is the point-present, the ever-sliding divider of time, as Aristotle had already noted, ever the same in kind and ever new in temporal location. Nothing could catch its character better than the following immortal exchange:

> A friend asked Yogi Berra: "Do you know what time it is?" The Yogi: "You mean now?"

No one can tell what time it is *now* and no one wants to know what time it was *then*. And yet we do tell time, in a badly compromised catch-up way.

Besides this inexorable directional flow, the Now is afflicted with all the problems attaching to an infinitesimal element of a continuum, problems that were so eloquently set out, shortly after the Calculus first came on the scene, by Bishop Berkeley in his *Analyst: Or, a Discourse Addressed to an Infidel Mathematician* (1734). The Now is a nothing that counts for something and a something that vanishes on inspection. Of such "evanescent increments" he says that "They are neither finite quantities, nor quantities infinitely small, nor yet nothing. May we not call them the ghosts of departed quantities?" (35). The Now is intelligible only as an abstraction, but as an abstraction it is no longer what it was meant to be in life.

(b) "Instant" usually refers to the physically filled Now and our perception and clocking of it, as in "the instant of arrival." Plato gives a first description of it under the name of *exaiphnes*, "on a sudden." He articulates the problem of the physical instant, that it is both time-deprived and action-filled:

> It is a certain strange nature, seated between motion and rest and in no time. And in it, and also out of it, the mobile changes into resting and what stands still into moving. (*Parmenides* 156 d; p. 5)

Hume resolves the question by opting resolutely for infinitesimals—indivisible instants of time (which he calls "moments").

> 'Tis certain then, that time, as it exists, must be compos'd of indivisible moments. For if in time we could never arrive at an end of division, and if each moment, as it succeeds another, were not perfectly single and indivisible, there would be an infinite number of coexisting moments, or parts of time; which I believe will be allow'd to be an arrant contradiction. . . . [These] indivisible moments of time must be fill'd with some real object or existence. . . . (*A Treatise on Human Nature*, 1739, I ii 2–3)

One huge consequence of positing such absolute perception-particles will be the loss of personal identity; there will be no way to collect these instants into a temporally extended unitary existence. The notion of the specious present, intertwining past and future, was devised to overcome such fragmentation (p. 142).

The instant of perception certainly has a time-extended biological background. The clock-time needed for a sensation-event to elicit a reaction is, by one reckoning, 9/50 of a second; there are various other estimates of the interval it takes for the subserving organ to process the sensory datum (Campbell, 230; Fagg, 172). If the "specious present" has a measuring motion, that is to say, a clocked duration, one of these might be its unit—or perhaps it is the three-second quantum that our brains are said to integrate into a chunk of present (Poeppel, passim; Gallagher, 161).

It is in the instant of consciousness that space and time, Here and Now, come together in "event-particles." Hegel begins with space and from it derives time and consciousness; thus space and time are dialectically the same, for the Now is only space differentiated, made to show glimmers of life (p. 26). A recent attempt to come to grips with this wonderful symbiosis has been made by Irwin Lieb. He begins with "individuals," self-reconstituting temporal beings, and so he begins with time; space is then the spread of time outward from these active individuals, arising from their present expansive activity. Hence space is the same as time, and it exists only in the present. "There is no space in the past and none in the future. Space is not an eternal background over which time passes and on which motions . . . are located" (Lieb, 63 ff.). Thus what I term the instant, when the human present, that is, perceptual consciousness, and the spatial infinitesimal, that is, a quantum of worldly variation, come together, is explained as a momentary spatialization of the temporal being. Whether or not this denial of the Gods' world alluded to above (p. 11) is quite intelligible, it is a serious attempt to explain the unity of temporal consciousness and spatial world in the instant. The problem has a version in physics: How can space have a ratio to time (pp. 17, 190)?

(c) "Present" is from the Latin for "being before, being in view." It is the instant regarded as a human experience. It is, as was said above, the central phase of time, when the world is before us, when newness emerges and action is initiated. It is the phase of our existence, "the locus of reality," and the time of action—the pragmatic moment (Mead, 1 ff.).

The very name and force of the present raises the question, already implicit in Aristotle's theory of time (p. 46), whether and how an empty present is possible. It is certain that there are powerful and varied experiences of mere time, of a mere, melded passage of Nows:

> Nor was there anything
> To make a daily meaning for her life . . .
> But the blank taste of time.
> Edwin Arlington Robinson

And in the opposite rushing beat:

> Time is a horse that runs in the heart, a horse
> Without a rider on a road at night.
> The mind sits listening and hears it pass.
> Wallace Stevens, "The Pure Good of Theory"

But whatever the tempo of bare time, it seems usually to assume the character of discreteness. All the forms of "doing time," of mere waiting it out and wasting time, of managing to endure and to accumulate duration, display time as a pulsing or beating, often mirrored and expressed as the smallest clock-motion in daily use, the second.

So Baudelaire, under the regimen of that nihilating melancholy known as "spleen," watches the clock, the "sinister god":

> Three-thousand six-hundred times per hour, the Second
> Whispers: Remember!—
>
> "The Clock"

So under the regimen of fear in a tyranny, Nadezhda Mandelstam reports in *Hope Abandoned*, subjects are robbed not only of memory, but of time itself; she means the carefree, unnoticed duration of a safe life. Instead "you begin to have a special awareness of each minute, each second" (251). Time acquires drag and weight in the small, while flying by shapelessly in the large. But so also under the regimen of love is heard "the colloquial music of time, which is only desire expressed in heartbeats"—pulses that Lawrence Durrell in his *Alexandria Quartet* learns to fuse into "that adventive moment which can't be measured" by "beginning to live between the ticks of the clock" (Durrell, 6). I will try to say later how a pure present, devoid of anything but consciousness, might be possible (p. 206).

(d) "Moment" is from *movimentum*, the Latin word for that single action that causes a turning of the scale. Santayana calls it the "conventional moment" to distinguish it from the natural instant; it is the ultimate phase, not always successively ordered, into which "sentimental time," time from the human perspective, is resolved (Helm, 91–93). It is the present insofar as it has force to turn life or make it significant. Thus Borges says in "A New Refutation of Time" that temporal "tautologies," the repetition of remembered moments, make up his entire life (*Labyrinths,* 23–24).

The moment turns up over and over in experiential musings—be it in prose or in poetry—about its import and its recollection:

> But only in time can the moment in the rose-garden
> . . .
> Be remembered; involved with past and future.
> And through time is time conquered.
>
> T. S. Eliot, "Burnt Norton," II

The momentousness of the moment, the poem implies, would be inoperative were the moment not seated in time, which acts both as bridge and as buffer—allowing us in time to recover a moment that has gained momentousness through time.

Nevertheless, the moment, though it needs time around it, fits uneasily into normal duration:

> Past, present, or future—which is the true time of man, in which is his kingdom? And if his kingdom is in the present, how can the now, by nature explosive and orgiastic, be inserted into historical time? (Octavio Paz, "Olympics and Tlatelolco," *The Labyrinth of Solitude)*

The integrability of a momentous present into the flow of merely ordinary eventfulness seems to me to be the test of its soundness; there are bacchanalian moments that blank out the ordinary time about them and bar the way back to normal life.

(e) *Kairos* and *Augenblick* are moments specially defined. Our absolute and universal chronology takes its "epoch," its point of origin, from a *kairos*, the momentous, brief irruption of divinity into duration, of God into space and time. *Kairos* was an ordinary Greek word for the right and opportune time, adopted in the New Testament and later in theology for the moment of the advent of the Christ (p. 193). It is the external moment of eventfulness, of an epiphany that reorganizes human time into B.C. and A.D., "Before Christ" and "In the Year of the Lord."

The *Augenblick*, the German word for the glance-and-blink of the eye, designates an intensely personal experience, when the receptive glance alights on a vision that presents itself in the world. It is also Heidegger's term for the moment when human Being, *Dasein*, is authentically, fully, placed in its situation, when it is there in "anticipatory resoluteness"; it is the *ekstasis* of the present (p. 106). But *Augenblick* appears in scores of other contexts with their distinctive descriptions: in theology, esthetics, politics.—There is a whole book devoted to it: *Augenblick und Zeitpunkt* (Thomsen and Holländer, 1984). As the moment emphasizes gravity, so the *Augenblick*, which has a perfectly ordinary usage as "instant" or "moment" in German, conveys at once the notion of a minimal of time marked out by a least perceptual interval, and of the irruption into life's mere duration of a singular, motionless epiphany. Such a moment reveals what life would be were it livable as a series of presents and presences rather than as an absentminded succession of mere melded Nows.

Perhaps *exaiphnes*, "on a sudden," should here be mentioned once more, for in some contexts it means the abrupt illumination that translates throught to a higher realm, the specific moment of the human union with Being (Plato, "Seventh Letter," 341 c, p. 100).

II. Time-Pathologies: Phase-Fixations

The reflections that follow might seem a little out of line for a philosophical inquiry into time, because they are concerned with matters of temporal sensibility rather than with temporal Being: how we do and should feel about the phases, what role we do and ought to assign to them in our lives, what their relative public standing is and should be. These are questions addressed more to experience than to analysis, and the answers are more directed toward recommending conduct than toward giving intellectual satisfaction. For it has seemed to me all along that the inquiry into time must have practical effects on the economy of life and the choice of ends. Thought about time cannot help being, as I said above, a proto-ethics. Not only are the fundamental intellectual choices that we make—

choices presumably based on comtemplatively dispassionate distinctions—an expression of what we think we can live with, but everything that comes into view thereafter is informed by considerations of good and bad. That seems to me to hold of all philosophical inquiry but most of all when the subject is time. For time, I will say along with quite a few others, is our distinctive human capability—together with image recognition and speech.

By a "pathology" of time I do not, of course, mean clinical illnesses, such as the amnesias and their equally terrifying antithesis, the compulsion of remembrance (Sacks, 153 ff.), nor the cognitive temporal scrambling that confuses inferential reason in schizophrenia (Mo, 247). I mean rather perfectly normal opinions and sentiments about time, quite commonly held, which may, although they are not in the least insane, look so unsound from the vantage of a contemplative armchair or from the perspective of an office charged with ministering to a community of learning as to seem more than a little mad.

Here my own sensibilities come into play, which are, of course, alive to our contemporary situation. Were I, for instance, writing in the Victorian period, when there was current what Tennyson called "The Passion of the Past," sometimes degenerating to a maudlin nostalgia, I would probably be inveighing against memorial luxuriating (Buckley, 111). But in our times the prevailing mood tends just the other way: It is the future that dominates our working discourse, with the present a close second when it comes to our leisure, which we curiously call our "time off." The public past, on the other hand, is sometimes revised downward by its historians, sometimes meticulously reconstructed in precision-built preserves by its curators, and mostly just ignored by people at large. So I shall be, cautiously, defending a rehabilitation of certain modes of pastness. But my passion will go into attacking the widespread current self-enslavement to the future. I see something persuasive in the nasty wisdom of a certain minor devil, who says that

> nearly all vices are rooted in the Future. Gratitude looks to the Past and love to the Present; fear, avarice, lust, and ambition look ahead. (C. S. Lewis, *The Screwtape Letters)*

Nevertheless, most of what I have to say is to be found in the vast, soberly secular literature on modernity, on our own age (of which postmodernism is a phase, perhaps not even a very late one). From my present perspective, the two defining features of modernity are those succinctly expressed by Milan Kundera in the dictionary of key words included in *The Art of the Novel*: "Glorification of the technical, fascination with the future," which features, he says, are accepted by certain modern artists with "*lyrical* ecstasy."

In this section, then, I will recapitulate the phases in a critical mode, beginning where I just left off, with the present, but now delineating its perversion into mere nowness, I will go on to what I think of as the chief of contemporary time-pathologies, the fixation on the future and the compulsion—as I put it to myself—

to form reception committees for a phase that, as reflection has shown, is not in its nature such as to be *received* but rather as to be *projected*. Next I shall go to the least acute of current time-diseases, though possibly the most chronically debilitating, the oblivion of the past, on purpose or by inadvertency. And finally, I will tell of an atemporal past, a phase different from those reviewed so far, which has certain curative characteristics: *aevum*, the time of the angels and of fictions.

1. *The Strutting Point: The Preoccupying Now*

"The strutting point," "the proudly exclusive Now," is Heidegger's rendition of Hegel's view of the present (*Being and Time,* 430). The point-now struts its stuff vis-à-vis all the other points in space, and that is indeed what makes a spatial point a Now for Hegel; it is a subjectively marked location in space. A Here-and-Now localized in space *is* a Now, and a Now is an elevated oversight of space that brings an incipient subjectivity into its dead expanse (p. 26). That is the dialectical view, the one that makes time out of space. Another, opposed perspective is that of Timaeus: The true present "is" not rightly time at all but a reflection in speech of that atemporal Being that is involved in the soul and within which is the place of the world (37 e).

Thus, however seen, the present is the break-point of time, when divinity, subjectivity, or human existence mark a divide in the temporal band. How could this moment not be a vulnerable spot in human temporal sensibility? What holds true of the present will be the case for the other phases: The vice peculiar to them lies, as in the scheme of Aristotle's *Nichomachean Ethics*, in two extremes, one of which is usually the worse (1109 a).

Thus what has been talked of since Max Weber first gave its name to a book as "The Protestant Ethic" (1904), that perverse but productive, religiously tinged mortgaging of the present to the future, may be an oppressively dour virtue, but it works. It is opposed to the spontaneity, that curiously self-contradictory plan for living in the present, which became a shibboleth of the sixties, and was soon degraded into the "now generation" of advertising hype, for whom the ideal of spaced-out *dolce-far-niente* turned into hectic recreational overstimulation. This temporal pointillism, devoted to instantaneous stimulation and associated with lifestyles rather than with lives, is mostly characteristic of the young (or of those reclaiming their youth) and eventually merges into the long duration of maturity; it is scarcely worth inveighing against.

Still, it is characteristic of a deeper temporal condition. We live in an epoch—of which, as I said, "postmodernism" is for present purposes an exacerbated continuation—whose inhabitants call themselves "modern." *Modernus* means "just now," and modernity is the epoch of constitutional nowness, newness. (*Modernus* was *not*, incidentally, used in the sense of "mode," fashion.) The use of the word to designate contemporaneity as opposed to older ages goes back to late antiquity. In the fifteenth century a battle, the mother of the succeeding battles between ancients and moderns, raged in the German universities; it was a fight between the "antique way" and the "modern way," roughly corresponding to realism and nominalism.—Nominalism is indeed "modern" in being a recognizable preparation for the advent of modern science in its emphasis on experience and individual judgment. In the seventeenth century "our time," contemporaneity, had turned into the "new age." Modernity meant newness, not only as a time of new beginnings but as ever productive of newness. But it was Hegel who first turned modernity into an epoch, our epoch (Curtius, 259; Coleman, 541 ff.; Habermas, 5 ff.; Koselleck, 231 ff.).

The tempo of modernity is, as one would expect, speed. If novelty becomes an object of devotion, then the faster the motion of life, the more novelty. For the Now to retain its standing as "just now," duration has to be decomposed into ever more novel Nows or, to put it simply: A lot has to happen all the time, and if eventfulness fails—for the eventful moment requires a quiet temporal environment—then at least there has to be mere speed, and eventually speed that far surpasses the human ability to experience it, as do the time frames of computers. Jeremy Rifkin sees in this disconnection of the tempo of computers and of the human beings whose mental processes they are supposed to assist a kind of temporal schizophrenia: The discernible human present vanishes into the compressed duration of computer-time, especially for children (Rifkin, 25 ff.). But perhaps this absorption in a world in which the specious present, the minimal duration of human experience, has been well and truly shrunk to an indiscernible now-point, is not so much a systemic pathology as a personal preference. In any case, for me the mischief some of our opinion makers make with the future seems more socially worrisome than various youthful preoccupations with the Now.

Still, Robert Grudin's warning is worth citing: "devotion to the present deracinates and destroys the present" (36). In the terms of this book: devotion to the *Now* destroys the present *moment*.

2. *The Slouching Beast: The Oncoming Future*

In hurried and harried times, time seems to come from ahead, as an assault. It comes, moreover, as an apparently irresistible superpersonal force, whether necessitated by the "logic of history" or fated by a divinity gone demonic:

> And what rough beast, its hour come round at last,
> Slouches towards Bethlehem to be born?
> Yeats, "The Second Coming"

But besides, or rather superseding Yeats's vision of ominous historical gyrations, there is a cheerful linear progression, secular and benign, but just as compelling. This future does not slouch, it runs, and it does not threaten the terrifying end of an epoch but promises limitless opportunity. It is not beastlike but machinelike.

But it is emphatically not the technology that is morbid here, it being the mark of sanity to be able to treat machines as value-neutral tools; it is rather the cheerful obeisance made toward this putatively oncoming future. Tocqueville had long ago observed that the democratic American imagination has no interest in the aristocratic past nor in the insignificant egalitarian present but is devoted to visions of the future (*Democracy in America* II, 1, 17). This future-mindedness has now become compulsive in some quarters. It is our curious case that just when the intellectuals have concluded that faith in continuous linear progress is misplaced, the managers of the world have entered a new stage of future-mindedness. And this future is not just desirable, it is also, they say, unavoidable, inevitable: Anyone condemned by responsibilities of office to read the position papers, thought pieces, roundtable discussions, administrative memoranda, commission reports, and all the rest of what the British so nicely call "bumf," knows the iron-fist-in-velvet-glove litany that goes like this: "We live in a time of rapid change. We live in the Information Age. Prepare to meet the future or be overwhelmed by it." The benign warners have neither religious nor ideological fervor but the comfortably positive virtuousness of those who know their time and conform to it; they are unexceptionably orthodox.

Despite my respect for Hegelian dialectic, I cannot understand Hegelian history, that is, logically, suprapersonally necessitated historical processes and prescriptively functioning epochs: That there is accelerated alteration all around me and that a stupendous amount of data is accessible at this juncture of affairs I would be a fool to deny. But that this would not be the case if my contemporaries did not severally put their effort into making it so, I must surely believe. So why should I go along with living "*in* a time," "*in* an epoch" named after quantitative growth that I know to be merely the resultant vector of myriads of human wills whose direction is not always determined by a human wisdom proportionate to human know-how? "Time" and "epoch" may be summary ways of naming what is now most talked about and is now most prominently done, but they are not the names of a temporal dwelling, a fate or destiny to be accepted. Until there are deeply reflected arguments why change *in general* (not this or that particular change) is a human good, why a rate of technological innovation that exceeds expectation and hence desire is desirable (Nowotny, 209), and how vast quantities of information (not just a pertinent selection) make for increased wisdom, my time-conscience bids me keep my distance, and so ought everyone's.

And most of all this is true of the advice to "Face the Future!" On analysis, the future appears not to be a phase out of which anything *can* come to be faced. It is not an oncoming potentate to be triumphantly escorted in but just an as-yet-powerless present to be actualized by that due balancing of probability and desire by which we recognize a serious intention of the will.

Today's situation seems to me no different. The so-called Future, with a capital F, is actually no one's fate but someone's present plan with which the rest are bidden to go along. Complaints about this Future would be mere kvetching, were not these plans projected as inevitabilities, so that fear of being overwhelmed and left behind for dead rather than reflection on what is truly desirable governs decisions—mostly institutional decisions. Whoever has watched, to give an example familiar to me, the managerial planning of educational institutions in this country and the courage called for in resisting future-driven changes will know with full concreteness the dangers I mean: being cowed by the threats and promises of a rampant futurity into betraying humane traditions.

But above all, the trouble is that the human happiness that is to come from all this is as indeterminate in date and form as the future itself, while in its pursuit the present is strongly burdened:

> Whoever builds a house for future happiness builds a prison for the present. (Octavio Paz, "The Other Mexico," *The Labyrinth of Solitude*, 283)

3. *The Night of Time: The Dead Past*

> I toward the night of time decline;
> You mount into morn.

So says Robert Louis Stevenson, contrasting England with America. Tennyson's "Passion of the Past" has already been cited. It is still true that a pathological devotion to the past does not affect us much in this future-directed country. The Romantic escape into the medievalism of the Pre-Raphaelites and the eerie embracing of the personal past that Charles Dickens describes in *Great Expectations*, where a cadaverous old Miss Havisham has wilfully entombed herself in the bridal clothes of her miscarried marriage, proudly nursing her broken heart, with clock and watch set permanently to a fatal quarter of nine—these belong to England and, in America, to Anglophile enclaves like Edith Wharton's Old New York or Savannah of the South. Pastoral nostalgia, rigidified convention, aesthetic inertia, were never characteristic of America, and, except for the ever incipient sclerosis of our bureaucracies, it would be a vast understatement to say that the constraining past is not now our main problem—though people stuck in that critique routinely pretend that it is.

But the opposite tendency has reached pathological proportions, and the discarding of the past proceeds in a multitude of concurrent ways: the rational eradication of such settled prejudices as, Edmund Burke thinks, form the wisdom of traditional communities (183); the planned and accepted obsolescence by which aging objects and ways are cast onto the proverbial ash heap of history to make room for the products of innovative progress; the radical revisionism of intellectuals that exposes the time-honored founding heroes to dishonor; the "sedimenting" of the root of scientific and mathematical discoveries observed by Husserl that occults the radicality of their origin (p. 154); the neutralization of the past by an academically passé but popularly bred-in-the-bone historicism, the doctrine that past thought is in effect *not* now present; and above all the methodical elimination of public memory practiced in tyrannies and its replacement by those works of propaganda called by Kundera in *The Book of Laughter and Forgetting* "the melancholy flowers of forgetting" (6, 1). These are some of the ways *external* memory has now or recently gotten obliterated for communities.

For the formation of an *inner* past, it is childhood that matters most, and whatever interferes with the slow, long rhythms of the imaginative life by which personal memory is stocked, such as a hectic and unstable home and continual flooding by indiscriminate external imagery, is apt to produce an insidious personal pathology with respect to the past.

There are certain trivializing ways currently in vogue to revive memory, chief among them *nostalgia*, the first time-pathology ever specifically diagnosed. Where long distances or fast changes separate people from their accustomed world, nostalgia, literally, "return-ache," arises. The word first appeared as a medical term in a treatise published in 1688, the earliest of many dealing with a disease, sometimes fatal, that afflicted exiles, particularly mercenaries—a genuine time-pathology. It is, as one would expect, endemic now in a mild, episodic, recreational form, a very personal, gently chronic sense of the age-old myth that time, by its power of decay, has effaced a Golden Age, the true human home, the "land of lost content," in A. E. Housman's words. It is to be indulged rather than assuaged in theme parks and period revivals. It is no longer a traumatic psychic illness but a soft sentiment that is agreed to be a fairly harmless relief from the losses of the speedy change, and perhaps even a way to tide us over the rage for forgetting. But it is a poor substitute for the vigor of a real renascence, which is not the desire to revisit the past on a sentimental holiday but to realize a vividly imagined archetypal pattern in the present, the public complement to the serious personal choice of a hero from history (Brandt, 59; Lowenthal, 4 ff.; Lears, 59 ff.).

4. *A Cure: The Atemporal Past*

What, one might ask, is the use of diagnosing ills without offering cures? Well, for the most part, the cure is just this very thing: inquiring into time itself, becoming aware of public and private mischief with the human phases, and then thinking and doing what is not generally thought and done.

One may read of all sorts of ways that human beings devise, besides like Shakespeare writing poetry, to "Make war upon this bloody tyrant Time" (Sonnet 16). They merge past into future and future into past by myths of an Eternal Return that like a vector sweeps out all the degrees of a cycle, while spiraling upwards, so that time's cycles are ever the same in quality though advancing in number. Thus everything both passes and is forever. Nietzsche says about the Eternal Return:

> To me . . . everything seems worth too much to be allowed to be so fleeting; I am searching for an eternity for everything. . . . My comfort is that everything that was is eternal—the sea washes it back again. (*Will to Power* 1065)

Or they diversify the mere qualitiless passage of secular time by sanctifying a day of the week as a reminder of the Creation: "Six days a week we live under the tyranny of things of space; on the Sabbath we try to become attuned to *holiness in time*" (Heschel, 10). Or they invest special spaces, sacralized landscapes, with long-lasting memory (Schama, passim). But there is a legion of ways to master our temporal vulnerabilities, too many to catalog here. I will add one more.

I want to introduce at this late moment yet another phase of time: *Aevum*, the time of the angels in paradise and of the imagination within us. It is, in its ordinary use, translated as "age," as in *medium aevum*, the Middle Age, but in its distinctive use it should rather be rendered as "agelessness." Eternity, sempiternity, "aeviternity"—these are the "times" beyond the human phases, the atemporal times spoken of in theology. Eternity is the atemporal Now; sempiternity is the atemporal whole of time; aeviternity is a lesser and a closer timelessness.

Now it is generally agreed by logicians and cognitive scientists and not just by certain philosophers that we have atemporal knowledge in our memory—knowledge of tenseless beings such as mathematical objects and universal grammar, elements that are intrinsically not time-tagged (though we may recall the time when we first became conscious of them), knowledge that has been "learned and not forgotten" in this life (Gilbert Ryle) or acquired genetically (Noam Chomsky), or recollected from a deep precongenital memory (Plato). I believe that it was Plato who first recorded that sense of dredging up from deep inside, of intentional remembering, that attaches to nonempirical learning and gave it mythical expression along with a practical demonstration (*Meno* 81). Ever since, going deep within has had a sense about it of going into the past, of passing through memory to beings aboriginally with the soul. Augustine describes this sense very vividly (p. 115). But *aevum* is *not* the time of this nontemporal knowledge.

It is the time, or better the temporality, of beings that have life but do not age. Like so many wonderful understandings of time, the formulation of *aevum* (though apparently not the word) originates with Augustine (*City of God* XII 15). One may find this involved notion named and debated by many scholastic writers, among them Robert Kilwardby in *On Time and Imagination*, written in the late 1250s (Questions 17–19), and very similarly a decade later by Thomas Aquinas in the *Summa Theologiae* (Question X, 5–6). Thomas places aeviternity between eternity and time, between highest heaven and earth. It is a lesser eternity, an edenic, paradisical time. Aeviternal beings, such as the created angels, have a beginning in creation but no end; they are unchangeable, since each angel is its own species, but they can have change "annexed" to them; they have no before-and-after, but can have time "joined" to them; they can assume incorruptible bodies not to serve their needs but to express their spirit, be it the martial goodness of a Michael or the rebellious badness of a Lucifer.

Moreover, aeviternal beings have no past or future, for these belong to flowing, human, time. They are neither in succession-time nor in phase-time, though these types of time are said to be compatible with them. And though these beings of paradise are immaterial, and so substantially immutable in substance, they are capable of change through the exercise of choice, and capable also of "acts of understanding, of affections, and of places, in their own degree."

Aevum is eventually absorbed into law, where it becomes the time attributed to a "fictitious person," such as a king, who is both a temporal and a perpetual being, "like the holy spirits and angels" (Kantorowicz, 171, 279 ff.).

Aevum is thus the temporality of fiction. Aeviternal beings are created, species-like, undecaying but somehow mobile, beings whose corporality has no function but to express their spirit—and that describes to perfection the beings that people fiction. Aeviternity seems to me to correspond closely to the mythical mode Vico calls "imaginative universals" in *The New Science* (1725, no. 209), the concrete universals of aesthetic theory: They have a maker and a beginning; for all their poignant particularity they stand for a human type; they act in their own time and as individuals, and yet they are as indestructible as universals. The persons and scenes of fiction are at once vividly individual and tellingly typical, in human time and out of it: "*Aevum*, you might say, is the time-order of novels" (Kermode, 72). And of course, of epics.

It is the cultivation of aeviternity that I recommend as a partial relief for our temporal ills. For the tyrannies of present, future, and past described above—preoccupation with the mere Now, submission to the oncoming future, and obliteration of the memorial past—are put out of commission in the aeviternity of works of art.

There are, then, many means, some of which I listed above, by which to defeat "this bloody tyrant, Time," and there are likewise several moods in which to attain timelessness. Just to give one pair of contrasting examples: There is the mood of the ascetic, world-abnegating melancholy of certain Christians in declining antiquity who, in seeking God's eternity, are looking for relief as well as salvation. And diametrically opposite, there is the lusty, world-defying vigor of those classical philosophers of Being who have a keen eye for their own brilliant times and a strong scorn for it as mere passing appearance. In our epoch it would be sheer temerity to recommend the timelessness either of God or of Being, but in the aeviternity of fiction we may at least experience the temporality of paradise in a permissible suspense of unbelief.

Chapter Two

Time: The Potent Nonentity

It seems to be in the nature of the elusive beast I am hunting that negative clues are more easily come by than positive propositions. For much that is written about time is an implicit—and sometimes a very explicit—deflection of the question what time is. You can see the tendency by comparing two etymologically parallel terms: *Ontology* tells what Being is; *chronology* lists events by date—the "account of Being" deals with Being; the "account of time" deals with dates. I will first work my way through some of those deflections, making a collection of things said about time that, I am persuaded, time is *not*.

And yet, it is perfectly plain that time is exceedingly potent, be it as an entity or as a nonentity. If it is a real power, it is the more potent for being ignored; if it is only a human mode, it gains power through the notion we have of it. I believe it to be the latter: It is a nonentity not, in scholastic terms, *omnino*, "in every way," but *secundum quid*, "in respect to something." I would put it this way: Time is so fundamental a human mode as on occasion to assume the guise of human Being itself. So finally, I will try to set out what, then, time *is*.

I. What Time Is Not

The first chapter of this book was given over to a preliminary clearing of the decks by showing that external, primarily physical, time was not so much nothing as a name for something else, in fact for quite a few things. For it never gives those evidences of existence required of other physical quantities like space or mass, such as effects caused or patterns traced. Of course, people are free to fall into any usage they please for purposes of brevity or poetry, and therefore they are free to call clock-readings, units of time from light-years to nanoseconds, the symbol "t," and directional processes, all alike time. The misrule of language rectification is always to be avoided, and in respect to time it would in any case be love's labor lost. My purpose in this particular critique is not to change the world but to interpret it. Moreover, for the purpose of inquiring into time it is good preparation to take people at their word. So below I review some of the ways time is spoken of and show what in these seems to me careless or unlikely or incoherent.

Time, then, is *not*: (a) constituent of external nature, (b) an abstraction from process, (c) a power or force, (d) a fungible substance, (e) a mere linguistic usage, (f) human Being simply, (g) an entity at all.

(a) External time. Most writers on the philosophy of science and most scientists, especially physicists, do, sometimes explicitly but usually implicitly, mean motion when they speak of time. To put that common opinion in relief, I want to cite a powerfully thoughtful objector to my view, Alfred O'Rahilly, in his *Electromagnetic Theory* of 1938 (vol. 2, ch. XIV: "The Symbols of Physics," 672 ff.). He attempts to save experiential time from those philosophical interpretations of Relativity that claim, in fact exactly as Einstein did, that time is only what the clock measures (p. 7). His objections are antecedent to any specific physical theory. He argues that these interpretations "belie our fundamental perceptions." They fail to distinguish between "real Duration and measure-number duration." Consequently they deal carelessly with the first dimensions of physics such as space and time and the ratio-relation formed from them, such as s/t, that is, velocity. Such relations, when misinterpreted as quotients, as divisions, become unintelligible to ordinary sense; what can we make of dividing a distance by a time (p. 17)? His solution is to be more meticulous: to regard the ratio *s*/*t* as pure algebra, which is given an interpretation only in the experimental laboratory. He is coming to grips with an epistemological problem already recognized by Descartes (Klein, 197 ff.), but it seems to me that he is pushing it off rather than resolving it. He says that Duration as a magnitude is unanalyzable: "So-and-so is happening *while* the hour hand is moving from 10 to 11. This is a simple fact of experience; it does not imply any reduction to clock-time, especially if the last is taken to mean a series of abstract numbers" (682). Here the burden of meaning is, of course, borne by the unforthcoming word "while." But who or what is whiling away [time] while the clock moves? Is the laboratory phenomenon enduring or is the scientist waiting? Until that question is answered the existence of "real Duration" has not been established, and the perceptibility of time as an external phenomenon, which is intended to be implied, is not proved.

But enough has in any case been said about external time in the first chapter of Part One, to allow me to reiterate now that "time" in physics is a misnomer for locomotion.

(b) Becoming. There are, however, in nature physical and biological alterations other than locomotions, the various processes. In these time might be found not as a measure but as a generator. For Aristotle it ought to have been, but failed to be, a question whether time requires change (p. 46; Sorabji, 67 ff.). But here the question is whether natural change requires time, or perhaps *is* time. Is some element, some feature of change, some pure Becoming, to be abstracted from the qualitative or quantitative variations of phenomena? Is what defenders of process-time variously call the vitality, the passage, the dynamics of the world just the process itself or the temporal aspect of it? Would one wish to argue that the natural processes themselves *are* time? The surely undesirable consequence would be that time and nature would be identical, since, at least when seen close up, nature is all change (though from a temporal distance, it has stopped durations, like the "punctuated equilibrium" of evolution and the quasi-permanent patterns of celestial orbits). If we conclude that time is an aspect of process rather than process itself, we should look for the mere Becoming in change. And such a feature is at least thinkable. But we must immediately ask: Is it insipid—a mere abstraction, or is it potent—a real mover? That is to ask: Is the time-aspect of change causal, and if so, in a formal or a deterministic or a probabilistic way (Savitt, 7 ff.)? If, as some have argued, time is objective and dynamic (Tooley, 377), if Becoming is a motor, what is left to do for the material element of change, its visible bodies and palpable forces with their qualities? Moreover, if pure Becoming is strong-featured, how can it be conceptually identical for all processes, so as to make time universal? Won't there be as many times as there are discernible processes?

I cannot see how these questions can be answered except by acknowledging Becoming as having the aspect of subjectivity. I don't mean subjectivity as Whitehead uses it in his theory of time: as the self-constituting of the episodes or event-quanta, whose coming to be and perishing defines the flux of becoming. I mean by subjectivity, consciousness (or the capacity for becoming conscious), just as does Hegel in his development of time from space (p. 26). I acknowledged above that Hegel's exposition of time as intuited Becoming, though unique in facing the problem of time in nature, seemed to me unacceptable both because of an ambiguity about who is doing the intuiting and because, as with all Hegelian concepts, the whole system is part of the bargain. But this part of his insight seems to me invaluable: that this pure dimension of change which is discerned as Becoming must arise from a meeting of subjectivity and nature. In the *History of Philosophy*, in the section on Heraclitus, Hegel speaks very plainly through his predecessor, who thought that "time is the abstract intuition of process . . . the first form of Becoming . . . pure Becoming as intuited. Time is pure alteration, the pure concept, the simple that is composed harmonically from absolute opposites." Yet it is this very composition of Being and Nonbeing that marks even the abstract structure of time as a human prerogative, for only human beings are able actually to experience and sometimes to articulate composites of Being and Nonbeing.

There is one possibility—which is no more than a devout wish—that time might be *in* the world: If the cosmos is, as Timaeus has it in his likeness-making myth, *within* soul and is encompassed by it, or if the world is, as Newton says, in a divine sensorium (p. 153). The world's motions would then be caused by, and present to, a divine awareness of which human beings would be the intramundial images. But as it is, pure Becoming cannot be time either, at least not as an aspect of external processes.

(c) Power. What if the purity were transformed into power, and the abstract structure of time were thickened into a moving force? Such a power would be externally causal. In nature it would be the agent of the laws of motion and process, in human affairs, of the laws of history and society, in the trans-human realms, of a demonic or divine will. Time is often spoken of in this way, causally. What could that mean, aside from mystical self-spooking?

The four causes established by Aristotle, four ways to assign responsibility for what is or happens, are a good guide to thinking through the notion of time as cause (*Physics* 184 b). The four causes may be represented spatially as working from below, above, before, and behind. The cause responsible as underlying is *material*; Newton did in fact think of time as a flux-like substrate in nature, but he was not persuasive (p. 14). The cause acting from above is *formal*; Kant's understanding of time as a transcendental form of sensibility runs into insoluble difficulties as well (p. 60). The cause coming from ahead is *attractive* as a purpose that pulls, but I think I have shown that futural time is not a compelling cause (p. 184). And fourth, there is the working or *efficient* cause that pushes from behind in the mundane manner of a shover.

This last causality, cause in the ordinary physical sense, is the most plausible candidate for causal time. Is time responsible for external change as an agency that is not the change itself but operates, forcelike, to effect it? A fantasy has been invented to show what that thought entails. It begins by imagining a way to separate time from the phenomenal change it might be causing (Shoemaker, 68 ff.). Suppose three regions, isolated from each other with respect to causal influence but mutually observable. Let one region go stone-cold so that all change there stops. When warmth returns and the observers who live there are unfrozen, they will infer from looking into the other regions that time has passed. Now suppose all three regions undergo differently phased cycles of deep freezing, and these cycles are established by their neighbors. Then all can together calculate when the freezes will coincide so that their whole common world is stopped dead and no change whatsoever occurs in any region. When they return to animation they can assert that clock-time has passed without change having taken place. But there is one little condition: Since all motion has (mysteriously) ceased, it will have to be started up over a temporal distance, as it were. Nothing is continuously active here but time, so time is the sole imaginable causal agent. We do not mind action through mere distance, but we seem to balk at action through empty time. Yet if we were to accept the possibility, we would indeed have achieved the notion of temporal causality apart from physical change. But there is a difficulty even with the mere thought: Action at a distance in space is among bodies or over fields of force that are three-dimensional, spatial configurations; it is not mere space that acts; it acts as being wherever bodies are and as patterned to produce effects on them. But how is unidimensional, unstructured time to be conceived as acting? How is one time to be the cause of many different changes and rates of changes? The fantasy fails because it is too fantastic: To concoct a setting in which time is observable in the absence of change, physical infeasibilities have to be imagined, while making it causative requires conceptual impossibilities.—Physical "duration causality" is unacceptable (Newton-Smith, 33 ff.).

The converse question might arise: Is there change in the absence of time? Atemporal change has certainly been conceived, and under various aspects. A number-sequence or series is normally thought of as a nontemporal progression.—Cohen thought he was doing something radical in introducing time into such mathematical structures (p. 91; Yourgrau, 161 ff.). Kant's manifold that gives matter to the form of time is, ipso facto, conceived as pretemporal. Mozart is said to have "heard" compositions atemporally all at once, and God's eternity is an event-plenum beyond time. The temptation is to see such timeless change by means of a spatialization, to represent it as a variety of configurations inscribed or located in simultaneous space, but that is surely an insufficient representation. In the best-known spatial formalization of atemporal Becoming, not an instantaneous space but something with more depth, as it were, is required: a world of four dimensions (p. 11).

I have dwelt on the atemporal variability of the spatial world because it seems to be a necessary consequence of the notion that time is not external, and in particular that it is not causal. As the consequence of a reasoned reflection it seems to me to deserve the status not of a mystification but of a mystery, and so it is seen by Bergson (p. 30).

The language of force is most often used of time in history. "The times" that time is now at are said to bring things about: "date causality" (Newton-Smith, 28 ff.). It is curious that talk of time-determined destiny is concurrent with the modern sense that the world can be changed by the rational will (Grant, 21 ff.). A closer look at the supposed force of historical time shows, it seems to me, that it is not "the times" we live in that are active, but the people that are passive. A large component of that bowdlerized version of Hegel's Spirit of the Times, to which compelling power is attributed, shows itself, upon reflection, to be what the Greek philosophers called "opinion," that curious human demi-capability of being verbally agile but inert in mind, of being oppositional in stance but conformist in faith—of having thoughts, but not one's own. The "ontologization of history" that makes a force of human drivenness is a characteristic of modernity, which, although it gives time such—as it seems to me, unwarranted—primacy, is, nevertheless, a study beyond this book (Galgan, 380).

An addendum: The power of time is often personified. Personifications are concrete pictures of strongly felt abstractions, and they reveal the temporal sensibilities of the epochs in which they flourish. In Greek antiquity, time as *Kairos*, the moment of opportunity, is young, bald behind, and only to be seized by the forelock. Characteristically, the biblical equivalent is verbally described by enumeration and not pictorially personified, as in the beautiful litany by the Preacher: "To everything there is a season and a time to every purpose under heaven: A time to . . . ," followed by twenty-eight opportune moments (Eccles. 3:1–8). Later on

"Father Time" grows old and has attributes of decay and destruction, such as an hourglass with the sands of time running out and a castrating scythe (Panofsky, ch. II: "Father Time," 69 ff.). In the late Renaissance particularly, people are becoming secularized, that is, time-obsessed in a passionately pessimistic vein. Time is the cause of everything and mostly a power of evil. It is not yet a historical force but acts personally. Thus time occurs in thirteen of Shakespeare's sonnets, and occupies twenty-two stanzas of "The Rape of Lucrece":

> Mis-shapen Time, copesmate of ugly Night,
> Swift subtle post, carrier of grisly care,
> Eater of youth, false slave to false delight,
> Base watch of woes, sin's pack-horse, virtue's snare;
> Thou nursest all, and murderest all that are:
> O! hear me then, injurious, shifting Time,
> Be guilty of my death, since of my crime.
>
> Why hath thy servant Opportunity
> Betray'd the hours thou gav'st me to repose?
> Cancell'd my fortunes, and enchained me
> To endless date of never-ending woes?
> Time's office is to fine the hate of foes,
> To eat up errors by opinion bred,
> Not spend the dowry of a lawful bed.
>
> (lines 925 ff.)

This is time as what might be called passionately positive negativity.

(d) Fungible substance. To the upper classes, time is the enemy and their chief labor is to kill it, but for the rest of us, "Time and Money are almost synonymous," says Henry Fielding in "An Enquiry into the Causes of the Late Increase of Robbers" (1751). Money is the universal fungible, exchangeable substance, and time too is symbolically spoken of as a universal credit card: It is an asset that can, like money, be wasted, well or ill spent, saved, invested. It can even be negotiated in a Faustian bargain: an intensity of Nows for a sempiternity in hell.

The most acute analysis of the reason why money and time are mutually interchangeable seems to me to be given by Marx. He explains that it is not money that renders all commodities commensurable but, on the contrary, it is their "phenomenal" value, which is expressible in money, that make them so—but that value is the labor-*time* invested in them. The temporalization of labor, its measurement in clock-time and its payment in time-wages, first makes universal monetary valuation possible: Not time is money, but money is, deep down, time. But this saleable time is alienated life, "man lost to himself." Marx is quite evidently distinguishing here between clock-time and human time. Poor guide to political economy as he has turned out to be, he is a most acute provider of deep reflections on the human effects of supplanting the inner sense of time by non-natural standard motions and their negotiable valuations. He calls it the human enslavement to inorganic nature, a conversion of life-activity from the freedom, which is the end of human Being in its human *essence*, to a mere means of *existence* (Marx 1867, "Money and the Circulation of Commodities," ch. III 1, 106; Marx 1964, "Estranged Labor," 112 ff.). We may conclude that since fungible time is both externalized and degraded it cannot be what time truly is.

(e) Linguistic usage. That the question "What is time?" should be downgraded to a puzzlement about the "grammar," that is, the different usages in various contexts, of the word "time," is what Wittgenstein demands (p. 113). We are to stop looking for what time *is* and accept a variety of related linguistic usages: "Philosophy, as we use the word, is a fight against the fascination which forms of expression exert upon us." As it happens, the last thing I want to defeat is my wonder at the meaning of words, and, moreover, working linguists happen to be professionally devoted to the notion that time-speech has an objective reference and is about something, namely time. Time in grammar (as linguists use the word) is expressed as tense, and tense has "deictic reference," which means that it relates "utterance to the spatio-temporal coordinates of the act of utterance"—and so to real existence. Not all languages incorporate tense into the inflectional system of verbs as does English, but all probably have time-indicators of some sort, like particles and adverbs. Furthermore, though it is uncertain whether all languages have ways of making nondeictic temporal references, meaning references involving not the personal reference system but rather a public calendar- and clock-time, children learning English, at least, gain control of such systems only after they have mastered the use of tenses: "This would seem to indicate that the deictic frame of temporal reference is basic to language in a way that the non-deictic frame of reference is not" (Lyons, 636, 677 ff.). Here is the linguistic counterpart of my claim that phase-time is primary (p. 202).

What is illuminating is how neatly the distinctions and problems of the philosophical inquiry into time turn up in linguistics. There is above all the fundamental and most problematic distinction between dynamic phase-time and static succession-time (p. 161) so exactly mirrored in deictic and nondeictic language. There is the identification of "the temporal zero-point of the canonical situation of utterance" that is referred to by the adverb "now," which is analogous to the pivotal present in philosophy. There is the same interest in the relation of "here" to "now" and their asymmetries, and there are similar observations concerning the protean uses of the future tense that indicate the indeterminacy of that phase (Lyons, 682 ff., 817 ff.). To me these parallelisms imply that the experience of time itself has shaped the permanent structure of language, just as the analysis of time yields the temporary perchings of thought: Both language and thought are "about" the same (non)entity: time. Only philosophy starts ever anew—or should, while language conserves old insights.

But there is also direct testimony for the independence of reflective thought from the suggestions of language, and it comes from the verb "to be." In English, wariness about the relation of Being to Time is made easy; the original connection of the word "being" to growing, coming to be, has long since worn away (though it can be seen in the Greek word *physis*, which has the same root as "being"). Moreover, our tense-system is cobbled together from words with three different original meanings: "was" (staying put, dwelling), "is" (genuine, true, good), and "will be" (intention). In Greek, however, there is only one stem, **es*, and therewith the temptation to think that Being is tensed because its verb is inflected, so that the notion that to be is one activity with three temporal phases is linguistically reinforced. And yet it is explicitly resisted by Plato's Timaeus, who insists that "was" and "will be" are to be said of Becoming in time, and only "is" can be truly said of Being (37 b; p. 6). Of course, he is not the only one who uses language to get behind language. The durative and veridical, that is, the time-defeating and truth-telling senses of the verb "to be" are brought to the fore by other ancient philosophers to combat its common "locative," that is, corporeal-existential signification (Kahn 1966, 254). In other words, a powerful agenda can bend the language to its intention and deflect ordinary usage. Language can guide thought but it cannot constrain it.

(f) Human Being. The claim that human Being, *Dasein, is* Time or rather Temporality, Primordial Time, belongs both explicitly and unequivocally to Heidegger. "*Dasein* is time, time is temporal. *Dasein* is not time but Temporality" (Heidegger 1924, 20). Heidegger is giving an early, stark formulation of his central claim: Human Being, or rather each human being, is its own temporal ground; this Temporality is the condition of possibility for its being not *in* time but temporal: It "temporalizes" itself through three ways of going out of itself in the *ekstases* of future, past, and present, from which ordinary succession and now-time fall out (p. 103).

Because Heidegger is unique in his strong claim that human Being is Time, or rather Temporality, one is tempted to think that if one raises questions about his consequent analysis the claim itself is shaken. And there *are* questions from beginning to end: Is it finally an ontological necessity or just a human choice that puts possibility above actuality, so that the future has a built-in prerogative over the present? How is the activity of self-temporalization to be understood, and is it permissible to ask who or what is the being that performs it? What is the literal meaning of the prepositions that describe the time-creating *ekstases*, since "being towards," "being alongside," and "coming back to" are meant neither logically, nor spatially, and certainly not temporally (*Being and Time,* 323 ff.)? Is the condition of possibility called Temporality actually capable of grounding ordinary time? And does not the division of the *ekstases* into authentic and unauthentic, proper and improper temporality produce clefts that make a common humanity quite impossible (Hoffman, 34 ff.)?

Nevertheless, even if none of these queries could be satisfactorily answered, I would still do better to say on independent and positive grounds why the identification of human Being with time is not acceptable to me. The chief element of the answer is a belief about human nature. For Heidegger human finitude is ultimately expressed in the fact that existence *ends*. But it seems to me that our finitude is better sought in the fact that we *begin*, in our birth as part-natural beings grown, not thrown, into the world, whose temporal frame is not ground but consequence of our natural constitution.—We do not "temporalize" ourselves; we are born temporal. Of the eight theories of time studied here, four—those of Plotinus, Heidegger, Augustine, and Husserl—are "diasporatic" or "diastatic," meaning that the human soul has somehow come apart, that it extends beyond or outside itself insofar as it is temporal. But all the writers except Heidegger understand this metaphor as a given condition and make it intelligible by specifying a center: The present is the moment of consciousness—whether temporal or atemporal—in terms of which the stretching of the mind into present memory of the past or present expectation of the future becomes, if not totally comprehensible, at least suggestively descriptive: Our present mind is not a mere point but has dimensionality.

Of these four, only Husserl infers that human beings must be temporal, that is to say timelike and fluxlike even into their most basic subjectivity. Plotinus gives the self-temporalizing soul a nontemporal origin; Augustine's soul is at home in eternity; and Kant, to Heidegger's disappointment, deflects the possibility of an ultimate temporal root (p. 56). But Husserl, unlike Heidegger whose analyses of human existence must be understood as having in view a Temporality beyond time from the beginning, starts with a genuine phenomenology of ordinary time-experience and is driven into ontological time somewhat against his will; he speculates because he cannot help it. And since he does it as one absorbed in the problems of temporal experience, he comes on an unnameable mystery, the subject's ultimately atemporal temporality, the absolute temporal flux (p. 149).

The lesson to be learned from his honest labors is not to take one aspect of human nature, even a deep one, for the whole. Most of our experience is time-imbued, but sometimes time recedes—because we are so absorbed in some activity that duration is nullified, because we are mesmerized by the vision of a moment, an *Augenblick* (p. 180), because we are, albeit temporally, occupied with atemporal matter, as in a mathematical insight. Moreover, usually whenever we are aware of living through time, there is present also external space with its deliverances; space is the constitutional partner of time. And that space with its appearances manages to turn up within us, so that just as we project time into the world by counting its motions, so we introject space by imagining its figures. Thus there is more to our inmost experience than the strenuously authentic derivatives of temporality. We may infer, or at least I do, that there is more to our Being than Temporality.

(g) Entity. By an entity I mean a determinate being that reveals itself through appearances to perception, by its powers and effects to reason, and perhaps in its intuitable form to the intellect. Time is not perceived by the external senses, it has no discoverable effects in the world, and it is supremely elusive to insight. It is therefore a *non*entity, but one of undeniable potency, both as an inmost human experience and as an organizing projection upon the world.

∞ ∞ ∞

Before concluding the study of time with conjectures about its nature, it seems right to give a skeletal list of the chief aspects of my subject that have been left out of consideration—really the bulk of what has been written about it. My object is to give the reader some idea of the scope of time as a study and to intimate the excruciating abstemiousness that must govern any single treatment of it.

First among omissions are the copious and diverse imaginative fictions that contain reflections on time or illuminating derangements of the temporal order. Some of these are among the implicit sources of the book, and I want to mention especially the time-fictions of Jack Finney, particularly *Time and Again* (1970), which manage not to be contrivances but real novels in which the confounding dilemmas of time-travel are poignantly imagined (MacRae, 374).

Next come clocks, which bear the brunt of time-telling and whose development is so intimately linked to ways of living. From the monastery bell (German: *Glocke*) that tolled the hours for the divine offices, to the public tower-clock that made time a communal possession, and finally to the portable watch that regulates the private working day and is often felt to be the devil's device, the instrument for telling time, "the mother of machines," was both the expression and the cause of great changes in the time-sensibilities permeating Western societies. In fact, the most temporally sophisticated of non-Western civilizations were evidently not clock- but calendar-oriented, and with respect to them the problem of isolating the ever elusive element of time is even more delicate than it is for civilizations that have time-pieces. Thus, although the Mayans' "concept of time" has received a great deal of attention because of the complexity of the Mayan calendrical system, no specific Mayan word for time seems to be isolable; *kin*, the inclusive word said to mean "time," is also the word for "sun" and "day." One might ask whether the obsession with significant epochs, as opposed to daily time-telling, perhaps inhibits the development of a separable concept of time. Yet illuminating as these facts and problems are, neither the history of the use and the influence of clocks, nor the development of their construction, nor the origins of the divisions and units of time (of which the non-natural ones, the hour and the week, are most interesting because they have sacral origin), nor the intricacies of calendrical systems are as much as touched on here. For then this book would have been endless. Let a short list of authors serve instead (Aveni, 193; Boorstin, 72 ff.; Dohrn-van Rossum, 29 ff.; Hall, 111 ff.; Thompson, 352 ff.; Whitrow 1988, passim; Zerubavel, 5 ff.). As for non-Western, above all Chinese understandings of time, I have left them out from simple incompetence; there is no grasping of time without some sense of the language in which it is spoken of (Frazer and Haber, 1986).

Besides clocks and calendars, some of the chief treatments of time in different disciplines are omitted. The time-theories of physics, except for certain elementary (but fundamental) aspects, are scarcely touched on, nor is time in geology and evolution considered.

But the largest lack might appear to be the absence of the logic of time, which is sometimes called the topology of time, since it is mostly concerned with the abstract configurations that are thought to pattern time. Time might be conceived with respect to extent as boundless (that is, cyclical, as in an Eternal Return), infinite (without beginning or end, as in classical physics), finite (having beginning and end, as in the Creation and Last Judgment), or any combination thereof. It might be conceived with respect to internal constitution as continuous or quantized, with respect to organization as unique or branching; of the latter possibility Borges makes a haunting story in "The Garden of the Forking Paths" (*Ficciones*, 89). It might be represented as a flowing or a static substrate, as being unidirectional or reversible, as wholly determinate or partly open, as a causal or a noncausal order. All these aspects of time have their formalizations. And above all, of course, there are logical approaches to the question whether time is primarily tensed or tenseless, whether it is flowing phase-time or static succession-time. These very technical expositions are not totally absent here. But they have been absorbed where they seemed to illuminate the philosophical inquiry; very often they assume a prior discursive understanding of the concepts they aim to formalize, so that what they gain in rigor they lose in vigor (Le Poidevin, 149 ff.; Newton-Smith, 48 ff.; Smith 1994, 17 ff.; Tooley, 253 ff.).

And finally, as if to demonstrate that, if in the study of time formalism does not invariably reach substance, empiricism is almost always trifling, there is the burgeoning field of research and advice on efficient time-use. But though it is nugatory regarding insights, it is quite powerful in effect, since it aims at the complete commodification and bureaucratization of the time of our lives. Here is a sample project: "Time is a universal and unvarying currency, and time-use research proceeds from the premise that ultimate constraints on both individual and societal well-being are the availability of time and a set of factors that determine how effectively that time is spent." This field, too, has been bypassed—though with much better conscience.

Having said at some length what time is not, and what aspects of the subject have not been closely considered, it is time to say what, then, time seems to me to *be*.

II. What, Then, *Is* Time?

The answer I am about to attempt to the question that serves as the title of this book and of this last section shares the character of the conclusions to all other such inquiries: It gives to an ultimate enigma that articulation which the writer can best live with for the time being. It clears up apparently inadvertent—but usually revealing—confusions, and relocates perplexities to a better place. It does away, as far as possible, with intermediate mystifications and preserves, without embarrassment, the final mystery of the matter. It is hard for me to see how anyone who has studied even one profound, broad-gauge work on time, not to speak of several, can do more. For all such authors write both deeply and differently about time, and that is good warrant for thinking that here is a subject that derives part of its unending attraction from being beyond human comprehension; the other part comes from its utter familiarity.

Here then is my long-postponed answer, in nine parts and with such reasons as I was able to figure out, to the question "What, then, *is* time?"

(1) Time comes from *within*. By "within" I mean what is called, at different times and in different contexts, consciousness, subjectivity, internality—and anciently and simply, soul. "Within" is meant to embrace everything from mere sentient awareness to self-conscious mind, but here it is used primarily of the latter. The most public feature of internality is its privacy. No mind- or soul-reading device, from the now practically proscribed polygraph to current brain-imaging devices, has yet succeeded in externalizing the soul in its specificity. The fact and the type of psychic activity may become evident, but not the feel and the particularity of it. Perhaps real mind reading is ipso facto impossible, but that is more than I want to argue here.

At this date, at any rate, it is part of being a self (which is another term for interior being) to be inaccessible to merely external observation. Who, for example, can tell for certain from appearance whether a subject is ashamed of having done wrong or embarrassed by being thought to have done so? Experienced observers can, to be sure, get limited insight into the motion of another's soul from unguarded bodily indices, but for real revelations of other persons' interiors to take place, they have to cooperate candidly and strenuously. In short, they have to be willing to "utter," that is, to make "outward" their thoughts and feelings, and even then with the best of wills, the revelations are only partial. Since for adults both bodily and linguistic expression is largely a discretionary activity, it is a lucky fact that having oneself under control actually tends to improve self-expression, so that the most self-possessed people are often the least opaque. But that is an aside, and the point is that we have an inside, whence time arises. Nonetheless, it seems sensible to suppose that the interiors of the animate bodies that I recognize as being species-fellows are pretty nearly like mine, that empathy can go further even than communication, and that human beings have their interior nature separately in common, so to speak. In any case, that is my presupposition here, as it must be that of anyone who puts pencil to paper about human beings.

There is a human interior then, and within it time originates, for me and for others. My reasons for thinking so are first negative and then, positive: Time never shows up externally (p. 15). It is never there to be pointed at as an observable entity, as one might point to a part or a content of space. Nor is it a relation of real things, for such a relation is invariably defined in terms of "before and after," and whether that relation is understood logically or causally or spatially—or by oversight even temporally—it does not put time into the world, but rather it reduces time to logic or to force or to space or to a circularity. But in fact, the "before and after" is always apprehended by a *soul* for time to come about (p. 46). Whoever attempts to show time in the world invariably presents us with a motion that is performed in a reliably equable way by some mobile, be it an oscillating atom or an orbiting planet, against which standard motion another motion is to be measured. And without a human mind to mark the Nows and a consciousness to count up the units so marked and a human being to call the result a duration, natural time does not come about, nor does social time.

Clearly then, I am not so perverse as to deny that there is a world without, external to us and extended in its parts. In fact, this double-outsideness—externality and extensity—must be closely related. What is outward to the soul is outward in its own nature; the duality would show a way to seeing why it is a spatial world we move in as bodies and take in through our senses and confront as souls: Space is self-external and so outside us by its very constitution—outside of our interiority.

However that may be, the notion of time as interior more than allows, it very nearly requires, the complementary notion of a world that is both physical and social, filled with inanimate and animate bodies and their motions and processes. This world is thus full of variations, some of which have patent orders, others hidden but discoverable regularities, and yet others may be ultimately random.

I speak of "variations" for want of any other, sufficiently nescient term. For if it is true that time comes from the inside, and moreover, as I will want to say, that we are with the world in only one of our phases, and that the briefest one, the present, then we simply do not know what the world does when it is not under the regimen of our time, when we are or were, out of contact, individually or as a species (p. 165). Some say that in itself the world is a fixed and absolute world-space, over which we crawl, speaking figuratively, since our advance is along the temporal dimension of this absolute world (p. 173). Some have said that the world as artifact or as creation has its own worldly time because it is itself the interior of an encompassing soul or of the sensorium of God (p. 153). Our ordinary experience is that we can move through space and observe its varying displays, or we can sit still and see the contents of space go by and alter. But what these variations betoken, whether they are temporal movements in the mind of God or have some, to us incomprehensible, mode of nontemporal alteration, that is a topic for speculation and such resolution as good faith allows and human viability requires. Meanwhile, it seems best to say right out that if time is internal, external variations with their regularities and causes stand forth as mysteries—which it is better to acknowledge than to obscure by a manner of speaking. For "external time" *is* a manner of speaking, an unmeant and unmeanable phrase (p. 3).

The positive reason speaking for the internality of time is the experience called "internal time-consciousness." Even the most passing introspective effort reveals a sense of time, of temporal passing—sometimes of a seamless duration of accumulated passage, sometimes of time ticking away in a punctuated progress. Above all, however, I discern in myself three "dimensions," "modes," or, as I say in this book, "phases" of time: past, present, and future. And so does everyone else who is sound of mind and is asked to report these experiences, and we all seem to live in these phases with varying degrees of acuity, vividness, and panache—or fogginess, insensibility, and distraction.

(2) The phases of time appear to me as the fundamental human temporality, our primary modes of time. They belong to the most intimate part of our humanity, and are a participating source of our various ways of life. They seem to belong to our humanity more distinctly even than thought and speech, which can be at least conceived to belong to nonhuman, whether sub- or suprahuman beings. Phase-time appears to be so powerful a defining feature, that as a mark of human Being having one's own past, living consciously and sentiently in the present, and intending to shape a future must supersede our biological classification as *homo sapiens*. Were a dolphin to have signified these capabilities to us, not in the enthusiastic hyperbole of animal ethology but extensively and really, that individual would have to be admitted to humanity and granted the rights appertaining thereto; consequently everyone who wished to study the rest of that species would have to obtain a signed release from its representative.

It has been said that of the three phases the future is most inherently temporal because it is furthest from atemporality: The past is frozen fast; the present is not a time at all; the future alone is fluid, and only time will tell what it holds. While I see the point, especially with respect to the present, this view seems to me to mistake the past. Taken as the whole of all *facta* it is forever changing by addition up front and loss in back, as well as by the continual resituation of the parts, for the memory is more like a workshop than a freezer. To me the past is the most characteristically temporal of phases because it harbors what "has stood the test of time," what has passed and survived, and so is a testimonial to the passage of time. Next in order of temporal savor seems to be the present, because, although under rational analyses it contracts to a Now, which is no time at all, in lived experience it is most pungently the time of our life; it is the motor and crux of the phases that, even as it stands out from the other two, it quickens and holds together. But above all, it feeds them: It is the present of perceptions, the moment of experiences, the Now of urgencies. From this perspective, the future is *least* a time, for it is nothing but the present at work: present conjectures based on inertial tendencies, imaginations of the moment hopefully projected, and propositions with no assignable truth-value now asserted. But these are preferences, though not mere preferences, since the choice of a phase-dwelling seriously colors life (p. 180).

The present is surpassingly confounding, primarily because it is so problematically temporal. If one adheres to the rationally abstracted mathematical model, in which the present has shrunk to an evanescent point-Now that is effectively "the ghost of a departed quantity," one can talk oneself out of life altogether and reach the absurdity that we touch down on the here-and-now so lightly as to almost not exist at all, for there is no mathematically respectable way that our temporal life-line can compose itself from bare now-points (p. 176). If, on the other hand, we allow our present its due extent, as we surely must, other ineradicable difficulties appear, and we have exchanged an insupportable absurdity for an unfathomable mystery—a good bargain, to be sure, but a disquieting one. One way to set out the root-difficulty is to say that if the present is admitted to be "specious" (p. 142), meaning that it is a beguilingly deceptive quantity, and since it cannot be past up to a point and future beyond (because that would yield the very point-Now cutting a continuum that was just rejected), it must be past and future at once—and now the present is so specious as to be spurious.

Yet it is now that we are here. Now and only now can our attention, focused or wandering, be directed to locations in space and what is currently going on there; it is only in the present that we experience places and take possession of them. It is indeed undeniable that, just as the largest part of our speaking is not communication but internal self-articulation addressed only to ourselves, so the vast bulk of our experience is internal and concerned with objects that we represent to ourselves, our own objects of imagination, memory, and hope, which are nonexistent and unreal. Yet, remarkably, it is only our brief present that makes us be *all there*.

I must inject here that nonexistence and unreality are not meant in the least to be terms of derogation. We "exist," literally we "stand out," on the temporal cusp as *being here and now*. Our perceptions are of reality, literally of "thinghood," be it ensouled or inanimate, that *confronts us here and now*. Existence and its external complement, reality, are fine and strong states, but there is much more in our conscious life than existence and reality; there are acutely actual states of our being that take place beyond existence and reality. Among these are our past and future.

There is thus equally no denying that along with all the unrealities we are presently conscious of, dreams of things gone by and things to come, there is the present proper, the living Now, that tiny window onto the real world, the conduit of perceptual newness and its peculiar excitements. This present is, as was said, the dispensing point for the other phases. It sucks in the future and disgorges the past, or, alternately, it tugs at the past and shifts it into the future. Without it, these phases are vacant, and there are no times. Yet the performance of this function as center, conduit, and bridge, all physiological, psychological, and introspective evidence agrees, *takes time*, whether measured by clocks or experienced by inner sense (p. 177). Moreover, this somehow isolable phase is felt to be bonded by retentive strands into memory and by protentive fibers into anticipation, so that, while it does not simply resolve into the immediate past and future or simply conflate them, it is yet intertwined with them. This fluxlike strand, into which the present is immediately bound, is distinguishable from the more remote past and future that are copresent in a different type of temporality (p. 141 ff.).

But the greatest wonder about the present is, after all, *existence*, this often-mentioned capacity of ours for being here and now, both at once. The German word quite ordinarily used for human existence, *Dasein*, "there-being," captures this meaning. Our own Latinate word for the replete Now, "present," also signifies in its two complementary senses that to be now is to be on the spot, face to face with real presence. We announce that we are present at the proceedings by answering "Here!" at the roll call of a world that shows itself as really "being before" us, as *prae-sens*. And these deliverances of language itself seem to express our experience well enough not to be attributable to mere etymologizing, which is a philosopher's trick for making the language say more than its present speakers ever mean.

We have strangely *actual ways of being* in the other phases, but it is, I am saying, in the present that we *exist* and have the world before us as real. It is *now* that we are *with* the things that are in space. One may exalt the present or despise it, but how can one escape the wonder of this momentary contact we have with an outside? There seems to be no way to deliver us from this strange pointillism of our existence. Even when we see ourselves as bodies moving lawfully in space, as animals responding to an environment, as humans living in society, we have not explained away what our reflected experience tells us: that our consciousness of the world is always "now" for us and we are always "here" in it, and immediately after this Now is another, and the world is new.

It is an inescapable speculation that this fit of now to here and here to now, this singular meeting of time and space, is somehow adaptive, that it is an evolutionary result. Perhaps it might even be the ultimate example of what evolutionists now call "exaptation," the development of a capacity antecedent to its most striking use. Our ancestors might, by some miraculous stroke of luck, have found themselves endowed with just these temporal modes in a world whose variations are most safely perceived instantaneously—perhaps because "human kind cannot bear very much reality"—and whose passages are most advantageously preserved in past form, accessible to future-intention. But speculation aside, the problem of absorbing human temporality, a deep aspect of consciousness, into evolutionary theory, how to integrate individual internal activity into the external processes of biology—that is, as the honest and strenuous thrashings-about of writers on human evolution show, a problem of the greatest delicacy. For it means discerning in external developments reconstituted from spatially stratified or physically dated remains, which is to say, from temporally interpreted space and matter, the earliest appearance of that very capability that first made such interpretations possible.—It means inspecting a bone and recognizing time. That is hard enough to do for the archaeologist, who is at least looking at artifacts that proclaim something of the mind of the maker, but it is infinitely trickier for the paleontologist who has much poorer evidence of interiority. Yet a thoughtful expert speculation on the survival value of phase-time would be truly interesting. For the time being, however, our two modes of living, the one called by the Greeks *zoe*, our zoological duration, and the other, *bios*, the sum total of our moments of existence, are still quite disparate.

Existence, then, when we reflect on it is speciously pointillistic, as it is evanescently pointlike when we reason about it. But when we live it out, in the ordinary half-attentive way, it feels quite continuous. The Nows blur into duration, perhaps punctuated by startled glances at the clock. To be sure, in deep abstracted contemplation of an internal or external object, time stands still, presence becomes atemporal, and the clock has no measure for such an absorbed presence. So also in high "orgiastic" moments, in the temporality of worldly or psychic festivities, time becomes quantized by the heightened quality of the event and stands out from the ordinary surrounding duration as a distinctive and discrete momentousness. There is, however, yet another way of experiencing time besides the distraction that accumulates time as duration and the concentration that disjoins it into moments. There is the focussing of consciousness on its own pure, empty self. I do not mean the reflective effort by which thinking turns on itself, but the sometimes achieved but more often unbidden awareness of our inner life, the purest experience of mere self-consciousness, in which our existence, deprived of its external "Here," comes to the fore as nothing but an internal "Now."

(3) It is there, I think, that we come on the origin of *succession-time*. I should recall here that the term "succession-time" is sometimes applied to the fixed time-line in which events are statically situated in a tenseless sequence. This temporal mode is "nondeictic," meaning that it is without personal reference (p. 195). It is an abstraction based on the linguistic usage that distinguishes the vocabulary of "past-present-future" from that of "before-and-after." It is the latter set of terms that yields tenseless succession as a concept. But, of course, no one knows for sure whether the postulated notion of unequivocally ordered temporal location is realized in the world, how the totality of such a linear ordering is to be imagined, and above all, what an atemporal before-and-after ought to mean (p. 161).

What I am here thinking of instead is a *personal* succession-time. It too is discrete, though with another, more aboriginal discreteness than that of the perception-replete present, which is isolated by reason of the almost unfailing novelty of each sensory instant. When the present, that is, the specious present, is empty of perception, either because nothing is in fact present, or because we cannot focus on anything, there comes to the fore what is probably our most acute type of time-consciousness, that in which we sense ourselves: *the beat of time*.

If in projective moods, when we anticipate events in images and designs, the future seems to be coming at us and slipping backward into the past, in the pure experience of mere passing time, it comes up from behind and goes forward. In empty awareness, time displays, so to speak, its own proper direction: onward.

Whence, one might ask, does it come, this forward tread of time? One answer might be that it is an epiphenomenon of our neurophysiological nature and expresses the irreversible processes that subserve memory (p. 14). But this explanation runs into the same old difficulty: The direction of these processes is not discernible except against our sense of progressive time. Without our watching and noting and remembering, no natural process is temporal, whether reversible or irreversible, and I, for one, cannot see how an experience can be explained by that phenomenon for which it is the condition of interpretability. Hence the question how forward-directed succession-time comes about is still open and askable.

Succession-time comes to consciousness, as far as I can tell, most palpably in that experience of mordant emptiness called "doing time" when it is involuntary and "killing time" when it is self-imposed. In waiting it out and in wasting time, when life has no object but that of passage, of being gotten through, as is the case for prisoners (or so I imagine) and in the state of fear or of boredom, that burdensome nonentity appears as an inner pulsing or beating. Testimony and introspection are the source of this description (p. 178).

It seems quite possible that this beating of time occurs in tandem with our heart beat. I have not come across such an experiment, but I can imagine the design of one in which subjects are asked to withdraw from sensory intake, and to empty their minds as much as is possible, and then to count up the mere elapsing instants (not using the usual verbal tricks for pacing seconds), while their cardiac rates are monitored to see if there is a common unit. But the question here is not about the unit-beat of time but about its underlying directed punctuality, the very fact of its beating.

This punctuated passage can be counted. No, more: It *is* a rudimentary counting, insofar as counting begins as a beating out of units. But to make this mere reiteration take off, it has to become an accumulation; one by one the units are *added up*. The counting-numbers, the positive natural numbers, are in fact a mnemonic system for retaining the results of the count at each moment of the ordered directional accumulation. The trick is to use very few conventional names together with some simple rules to go on indefinitely naming sums without being obliged to retain in memory all the counting-moments that led up to them.

Now in fear or pain or sheer waiting, time beats on without any summing: episodic and uncollected, aboriginal and unbidden, through each tensely empty present, until the siren goes by the house or the pill goes to work or something finally happens. But the mere pulsing of time can be and sometimes is deliberately noted and counted, so that while the beats slip away, a time-count is established. And, of course, each counted moment can also be associated with the present stage of a perception or a feeling.

What I mean to describe is a succession-time that has two stages: a mere episodic beat that appears in an intensely denuded present, and an accumulated count that results from attentive time-noting and time-telling. In the first stage mere originary beating time occurs, in the second its directional order, temporal succession proper, is verbally articulated. We are counting our own consciousness.

(4) But who is really counting what? The counting that is internal time-telling is an aboriginal sort of psychic pacing that first comes to the fore when there is nothing to count. It is a *self-counting*, and it may well be the transcendental source of all object-counting. Certainly, those philosophers who make an effort to connect the soul or subject to natural motion come on some such originary psy-

chic counting (p. 35). What I am saying is that, whatever deep roots it may have, it is also an experience that can be made available to introspection on demand.

But this notion of an originary pulsing and a subsequent counting turns into the following perplexity: It presupposes "some sort" of psychic motion (pp. 46, 73). In nature we note and grasp a motion through the mobile that performs it. But the soul's motion includes no mobile, as the counting has no objects.

I pause here to point out once again that just as I postulated a world so I am presupposing a soul. The internality of time requires, and indeed implies, some sort of consciousness distinct from and even opposed to the spatial world. Whatever the constitution of internality may be—a place of forms, a cooperating structure of functions and receptivities, an intentional subjectivity, a self-constituting flux—it seems to me to be *other* than nature (p. 18).

In no other department of life do we permit ourselves to reduce one thing to another unless we can show a thoroughgoing identity of matter or an isomorphism of form. But read where you will on the subject, among the simpler reductionists consciousness and its particular contents—thoughts, images, perceptions, feelings—never seem to bear a satisfyingly recognizable relation (not even necessarily of resemblance) to the piece of nature to which they are being reduced. And where the reductionists are very sophisticated, the lay reader will have trouble telling whether finally mind has been reduced to matter or matter to mind. Consequently I feel entitled, on the basis of my own interior life and my fellow humans' reports, to posit a soul that sometimes initiates, sometimes alters the motions of nature, that has intentional and reflexive structures radically different from those possible to bodies (p. 129), and that is largely unamenable to physical inspection. For this is certain: Without the soul, time is a brute puzzle.

Yet wherever we read, we also find that the soul or consciousness are spoken of under the figures of space, body, particularly the human body, and the senses. These figures are legion in philosophy, and the key diagrams of this book are a prime example (pp. 123, 135). Common usage too is full of metaphors that are by no means dead: We feel "deeply moved," we are "filled with feeling," we "see" the truth of a proposition. The most poignant writer on time produces this amazing figure speaking of unbidden volunteer memories: "These I drive away with the hand of my heart from the face of my remembrance" (p. 115).

We speak metaphorically precisely because nonspatial motion is unintelligible to us directly. For a version of the old dictum that there is nothing *in* the intellect that was not first in the imagination seems to be operative here: Nothing can be said *of* the mind that was not shaped in the imagination. That is why philosophers so often engage in a manner of speech I think of as "allology," saying something other than what is meant; they regularly warn us not to *envision* what they so picturesquely say, but to *think* it.

Clearly here, too, talk of psychic motion, even if qualified as a "certain kind of motion," is figurative. Yet though the utterance be metaphoric, the experience is undeniable. I know no name for this motion without a mobile and displacement without change of place. There is no pretending to understand psychic movement, but there is no doubt that it occurs: The soul itself, or something "in" the soul, moves—and we are borne through time.

One specific and pertinent perplexity is whether a temporally active consciousness must be thought of as a flux. Should one speak of "the stream of consciousness," meaning the internal appearances that float in the temporal form, or must one say "the stream that *is* consciousness," meaning the temporal form itself? Yet how can a flux be ultimate? How can human Being be at bottom a Becoming? "For all this we have no names" (p. 152). Only this much is clear: One of our two kinds of time, succession-time, requires psychic motion for its generation, and it seems to be, at least at the origin, less like a continuous flux and more like a discrete progress.

(5) Succession-time has descendants, *derivative times*. By one of those dilations by which reason transcends the representative capacities of consciousness, we indefinitely extend and externalize the episodic counts. Thus we get the notion of a long-term world of which we occupy a tiny temporal segment. We attribute to nature indefinitely great numbers of unit-times from light years to nanoseconds, countable at all locations of space, those close to us and those at the edges of the universe. We imagine this time as both an indefinitely long series and as a universal web. All sorts of combinations have been thought of, from the unintelligible notion of a time that is altogether natural while we are mere observers of the clock, to the idea that nature arises within our temporal receptivity (p. 64). The extent of external time too varies from linearly or boundlessly infinite to finite at one or two termini. Both Christians and contemporary cosmologists speak of the beginning of time, the former as the acknowledged mystery of divine Creation, the latter in physical terms, that is, as self-creation. But it must be said that at least the popular expositions of cosmology are so equivocal (or perhaps confused) in their references to physically defined, humanly experienced, and philosophically analyzed time that it is unclear even to critics in the profession in what sense it is *time* that might be said to have a beginning. In any case, the present point was only to list the various sorts of extension attributed to natural time, and perhaps to note that these projections, all immensely large, cannot help but be reflected back to us and to shape our sense of ourselves as beings who have an evanescently small life-time but who reach in their science to the temporal limits of a mind-bogglingly long-lasting world (p. 17).

I have dwelt on the discreteness of aboriginal time-telling as a *fact* of experience. But I think it also makes sense on *reflection*. Since the intense moments of sensory life are distinct from the surrounding ordinary duration, why wouldn't their voided introspective counterparts be similarly discrete? They are both reflecting the structure of our capacity for noting attentively either perceptions or our own internal consciousness. I speculate in passing that the pointed nature of attention is related to the discontinuity of thought, which itself takes flight and perches continually (p. 160) and which always craves discreteness in its objects. For example, we intuit and represent spatial structures like lines continuously, but to master continuity rationally, we must become blind to the line and attend to the points on it, or to the numbers assigned to these. (Though why thinking has this affinity to counting it would be folly to get into here, except to speculate that our temporal constitution may be in play.)

But, of course, we can also make the converse move, and go from the points to the line. And that is exactly what seems to happen to counted time. We may begin with sporadic stretches of counted beats, we next extend them in thought indefinitely forward and backward, and finally we spatialize them to make a mathematical model of time that is analogous to the continuous character of external motion-paths (p. 40). It hardly matters whether this conversion is effected by thinking of the units as extended measures or by thinking of each now-mark as the limit of such a measure-stretch; in either case "those deserted spaces which time misses between the ticks of the clock" (p. 179) are filled in, and time is represented as a linear continuum. This model might even branch, if different time-lines are thought to apply in different regions of the world, so that time runs in parallel rather than in series. It might be universal or local. But whatever its topology, it is a continuum. This spatialized abstraction of time is encouraged by a common experience of internal time-consciousness: Just as when the inner noting of the beat lapses and it goes on only half-attended, there arises that sense of continuous passage I have called duration, so retrospectively, the counted accumulation blurs and melds, somewhat as in spatial perspective, where the items near the foreground are distinct, but blur into each other as they vanish into the background. So also the perception-filled Nows that were discrete in the present meld into each other as a duration-smudge in the more and more distant past (p. 31). Duration is, for authors who have no deep interest in time, its principal feature, and in one version it is said to arise from the gaps in our mental life, from the distances between one idea and another in the "train of ideas." This description seems to me to need modification (p. 65). Ideas don't have spaces between them; they are usually packed in a filler of mental debris, and duration comes about just when the ideas lose their original singularity and discreteness in retrospect, when they cease to be single acts commanding pointed attention and enter memory as part of the throng of unregarded has-beens of mental life. Thus duration, temporal distance, stretch-time, falls out from beat- or pulse- or succession-time.

So also succession-time itself had fallen out from phase-time, the fundamental temporal mode. For it comes to consciousness in the phase of the present, in which both its pulse is felt and the current time-count, which is a mnemonic of past counting, is alive. Thus personal succession-time is anchored in the present, but carries in it a reference to the past and a drive into the future.

(6) Yet phase-time and succession-time are also different at the root, as different as the two kinds of clock-dial in which they are externalized: The digital face

shows nothing but the counted time-number at this instant, the round analogue face shows the present moment surrounded by the half day that is no longer and not yet. In fact one might represent the difference between the two times as that between arithmetic and geometry. Before-and-after, or succession-time, arises by the bare internal production of numbers, while past and future require an internal receptacle or quasi-space such as can hold memory-images and future-visions. Succession-time is ever lapsing and continually driven forward, while phase-time is simultaneous, in the sense that it is all now, and every memory is reachable more or less immediately in memory-space, without any reverse recapitulation of the temporal succession, or with very little. Yet, oddly enough, succession-time produces a very elastic sense of duration whose *longeurs* are not measurable by the clock, while the memories of phase-time are pretty reliably fixed in their successions. Temporal length is more elusive than temporal sequence.

In sum, our sense of time has a *double root*: one is our ability to articulate consciousness in counting, the other is our capacity for perceiving presences and for storing absences.

(7) All the phases are *present*, contemporaneously present:

> Nor do we properly say: Three times are there—bygone, present, and future. But perhaps we do properly say: Three times are there—a present of bygones, a present of presences, a present of future things. (p. 120)

What have we done when we have reduced the three phases to three-in-one, and that trinitarian one is the present? But probably the better question to ask is: What else could we have done? To be present means to be there, *before* or *in* or *of* consciousness; it does not matter for the moment which metaphorical preposition we choose. So if to be present means to be before or in or of consciousness, the converse is close to hand: To be conscious usually means to be in the presence of something, and that "being in the presence of" is what in temporal terms is meant by *the* present—when consciousness comes into its own by connecting a Here to its Now, where the momentariness of consciousness is, for once, not the issue but rather that presence and the present belong to each other. It is, I should say here, no secret to me that "presence" is proscribed in contemporary continental philosophy, but this being America, it is possible to secede from this view. One conclusion of this book is that such an understanding of time as is possible for us begins with the present and its presences.

The main point, however, is that if we are compelled to think that presence and the present are the steady concomitants of everything consciousness is "about" or is "of" (to use the terms of intentionality, p. 129), then the remaining phases of past and future *must* be found under the aegis of the present.

And this can indeed happen, because present-consciousness, "the present" for short, is Janus-like and has, as do we, two faces. Toward the outside it is the phase of perception, of our existence with the world; it is the Now, the instant of the new. Facing within, it is the phase of internal consciousness, of our inner being; it is the copresence of all our times, bygone and yet to come. Of the two facets of our present, existence seems more acutely, more outstandingly, more excellently, the human condition. Philosophers have denied that existence is a "real predicate," one that adds anything to the concept of an object (p. 166). But in ordinary life existence is felt to be very much of the essence, and the object of one's desire really obtained is a changed object in a changed world: To dream as souls and wake as existences to find the dream true never betokens merely a change in our existential status, but rather a transfiguration of the world.

Yet when I say that existence in the face of reality, life in the outward-facing present, is most excellently human, I do not mean that it is best or most blessed, only that it is most poignant. This inward present, in which memories and hopes stand clear and free, not as a half-occluded background to action in the world, often seems a better, more spacious temporal place, even if it lacks the bonus of existence.—It is, of course, the clock, that false pretender to the office of timekeeper, that enforces our dutifully punctual involvement with reality and prevents a turn toward the inner present. But more of the inward face of the present below.

There are, to be sure, people who think they can regain the past and coexist with it right among the realities of the world. They think that the antique pot in its trench *is* a piece of the past, and so is the ancient fossil in its rock. But even the most time-rapt romantic will acknowledge on reflection that these objects are real *now* and share our present, and that we alone make these mute messengers speak, and that being old is *not* being past (p. 167). Even the most telling testimonials of the past, our public memorials and memories, are unmeaning presences without the back-projection of our private memories. And if there are people who go even farther and imagine that bygones still exist somewhere in the world, occulted and gone out of sight in some past-preserve there to be found, they are fantasizing, and such imaginings are more an airy amusement than a solid recreation of the spirit.

The past, we should rather say, is the sort of cause that sacrifices itself for its effect. It has disappeared into the present. Or, alternatively, we may say that things externally present *are* the past as we *are* our food, but that they *have* no past since they possess nothing bygone consciously. Hence our perceiving present is, once more, preeminent as presenting us with the sole survivor of the past: reality.

But the present presents us also with internal unrealities, whose less insistently pungent, one might even say, less confrontational mode, we might well learn to savor and respect. The primary phase of unreality, the past, will be taken up next.

But first to recapitulate: Anything that is in some sense for us and before us is present and in the present. And so the past is present, though not as real, that is to say, as external, as things in space, but as unreal similitudes within, as memory-images. Only we must not say of the past that it is composed of the sort of images nowadays talked of as "simulacra," images without originals of higher existential status than themselves. The images of memory do precisely have such originals, the real things or events, the *facta* that are bygone. I for one do not, incidentally, believe that the images that are *ficta*, the images of the imagination, are simulacra either, but I rather cling to the unfashionable view that they too have originals that are only "sortally" traceable to perception, that is as perceptional sorts or types; their fragrant particularity seems to come from another realm, the home of the nine Muses (p. 188).

Note that in these reflections it is the past phase only that I am attending to, because the future is constituted just as the past is—from images that come partly out of memory, partly out of imagination. The difference is that instead of the faith that the thing has been, there is the fear or the wish or the will that it shall be.

Note also that the idea of the cotemporaneity of the phases implies, or at least suggests, that we can on occasion perceive and imagine simultaneously. Some single-minded inspectors of consciousness have flatly denied this (p. 52), but even apart from the underground shaping of perceptions now generally conceded to memory and imagination, introspective experience seems to me to belie the denial: Much of the resonance and depth of our own perceptual life seems to depend on the double-seeing by which the bare present is invested with ghostly visions of memory and imagination.

From the perspective here pursued, the past is, then, not primarily a place, "a foreign country," or even a stretch of time, upon which events are strung. It is rather a set of discrete, singular images, identified as memories by their un-reality, here meaning their bygone reality; the passage of time has de-realized them, one might say. They are arranged in more or less well-ordered files. From the experience of their unanticipated return or their deliberate retrieval we infer that they are in storage; memory storage is a figure to express just this out-of-sight availability. The quasi spatiality of the imagination, however, is not altogether a metaphor, but rather a mystery, the subject of a previous book (p. xiii).

Yet if there is a sort of memory-space, the past is a place after all—call it Memory, whose singular contents are our memories. That admission corresponds to a familiar experience of the Past as a poignant, encompassing, sometimes transfiguring whole beyond and behind the no-longer-real facts that survive as similitudes. But if we let our imagination take us beyond the positive mundane model of a storage-space, another figure for our phase-temporality comes forth that displays the peculiar oscillating inversion so characteristic of representations of psychic dispositions. In one posture, the present encompasses the other phases and enfolds them as they fold in on it.—Every time is present. In another, these phases surround the present and stretch away from it, in an elongation of the temporal soul about the present.—The present is a pivot of times. The inverting figure, in which the present seems now to embrace and again to be embraced by the two other phases, like one of those Necker cubes that switches aspects under prolonged viewing, seems to me to represent pretty well our strangely structured temporality: centered, stretched, and stratified—centered on the here-and-now, stretched out into past and future, and stratified in a simultaneity of all three phases.

Of course, the two flanking phases are not actually in consciousness to any large extent at every moment, but they are felt to be ready to be brought back to the present. Perhaps the word "potential" describes their inactual presentness; it means no more here than that we impute to our memories an ever-present availability, if under modification, and a power to move us, even when submerged. "Subconscious" is fine too, if by it is meant merely that most memories and expectations are submerged by reason of the selectivity of our attention.

If our present is a triple-ply overlay and underlay of phases, memory is itself a sort of laminate. Memories are all of the past, but our past is closer in and farther off, and moreover, these positions change continually and modify the memories that occupy them.

Two "motions" seem to be at work, each having a more passive and a more active aspect. We increase the store of our memory continually and incrementally, usually without special attention, though sometimes we make an active effort to commit a perception to long-term memory (p. 171). It might be a face or a landscape we wish never to forget. Evidentially we are helped in remembering those elements of the world that are of particularly ardent interest to us by the "domain-specific" memory-modules for faces, places, and things, discovered in brain studies. But as we feed the memory-store up front, we incur a mainly passive slippage in the rear, a dimming reversed only by occasional reactivations. And this slurring of memory is just the connecting link between phase-time and succession-time described above (p. 209). The pulsing present is at once the phase of existence and the motor of time that drives the now-new Now out of short-term memory back into ever-deeper history; it is, moreover, the motor both of succession-time and of oblivion.

The other operation retrieves memories. It can be an effortful, methodical rifling through the memory file, as when we search for that lost key, first retracing our rounds in memory and there coming on it where it was last seen, and where it is, we hope, now in reality. That deliberate retrieval is called recollection—and, as was anciently observed, the pursuit of a priori knowledge has a similar feel (p. 187). But also all during the waking day, and even more powerfully during sleep, memories rise to the surface unbidden, fleetingly or insistently, though sometimes triggered by sensory catalysts, especially impulses from the nonrepresentational senses (p. 163). Many detailed descriptions of volunteer memories exist, but their antic associationism remains largely a mystery. Nonetheless, consciousness must be said to be strongly intentional with respect to memory, in the general sense that even random memories are contents of consciousness, and in the particular meaning in which, when we intend to become conscious of the memory, we shape for it a sort of negative receptacle into which it will seat itself, with a mental click. This is a figurative way of dealing with one aspect of the old problem of mental searching, which is first, how to target an unknown object, and then, how to recognize it when found. But here the subject is not the mechanism of memory so much as its temporality.

The sense we have of the temporal distance that a retrieved memory has from the present is the effect of the mental leap required to land on it directly, or the internal rifling needed to reach it by means of other memories. The memories between the present and the point in the past to be recalled form a sort of hurdle whose thickness gives us the sense of having leaped deep into the past. Alternatively, we sometimes seem to be going through a box of transparencies or a carousel of slides until we light on the one we mean to pull. Temporal distance, then, is a memory-laminate, whose thickness or number of lamellae gives us a rough sense of duration (p. 171).

But this pleasing picture fails before the variety of life's tempi. The sparseness of the memory-images that do arise in a long perceptual desert makes for a deceptive sense of brief durations, though watched clock and marked calendar may evidence time-without-end. Thus also the recurrent perception of a daily routine tends to merge identical memories, so that a third of a century may have the temporal feel of a day. Our sense of qualitative pastness is rarely confused, but our sense of quantitative duration as measured against external motion, that is, by the clock, is mostly unreliable. For memory-images are produced by the discrete moments of vivid perception; the *longeurs* of mere passage go undepicted.

The Nonbeing that marks memory and imagination shows up differently in them. The images of the imagination are not tethered to bygone fact, or set in our aboriginal time. Their originals were not in the real world, and their time, the ageless time of fiction, is parallel but not identical to that of our existence (pp. 146, 187). Memories, on the other hand, are essentially de-realized perceptions, and their order *is* the primary temporal order, that of bygone facts. Recall that by reality, I mean the objective correlative of existence.

We exist "here and now," and so the world is real for us. The past is therefore "there and then." For although abstractly considered, the same Here can endure many Nows, and a motion from here to there appears to be reversible (as a passage from now to then can never be), in fleshed-out truth you can't go home again.—A later Here, even if you are on the spot with identical coordinates, will be different by reason of the temporal distance and the nonstop variations of the world.

The consequence is that the kind of absence that characterizes the past which is now present for us infects not only perceptions and their moments but also their containing place. Hence whoever thinks that the unreal past is at least anchored in real places is banking on a rare phenomenon and one that requires a lot of imagination: the experience of widely separated Nows in identically preserved places.

And in general, memory and imagination, bygone facts and aeviternal fictions, are closely intermingled, since the factual memory, as it drops back, is almost inexorably transformed into a fictive image, while imagined images are often artfully assimilated to the factual past. So the question which of these two capacities for Nonbeing is primary and encompasses the other, that for imitating what never was real and that for imaging what is no longer real, is variously answered. The mouthpiece of one who should know speaks of his poetic gift as

> A foolish extravagant spirit, full of forms, figures, shapes, objects, ideas, apprehensions, motions, revolutions: these are begot in the ventricle of memory. . . .

So here memory is the mother of the Muses. Nonetheless, I conjecture that memory, which is the human capacity for holding images of unreality, that is, memories of past realities, is a specification—perhaps even one later in time in the species and the individual—of imagination, the primary ability to contain those conflations of Being and Nonbeing called images (pp. xii, 163).

(8) Does it make much sense to ask which phase is our "kingdom" (p. 179) when the point of the last pages has been to insist that they are present altogether in the present?—Do we want to distinguish one phase, when that bothersome problem of so-called "personal identity," the perplexity how we can know ourselves as one, how to remain connectedly ourselves through the slipping stretches of succession-time, seems to be alleviated by the thought that past, present, and future are all copresent to us? Then, too, time-trinitarians might on principle refuse preeminence to any one phase. And there might be those who think that our kingdom is *not temporal* at all.

I will plunge in and say that if a practical answer is to be made it will depend on "the times" (p. 181).—Whatever is the phase of the day most subjected to denigration or derangement, that is the one most in need of rehabilitation. I have argued that it belongs to modern times to be deranged about the future, that the pressure to anticipate The Future "proactively" is often a passive acceptance of unexamined purposes, and that this future-madness entails a massive carelessness or even hostility to the past (p. 185). Therefore the past needs to be reclaimed not least because from it come those fleshed out, vividly detailed images by means of which the weirdly virtual oncoming future prognosticated for us can be overcome.

Although this book has been about human temporality, about existence, reality and the phases of their absences, there are moments enough when we lose the beat of time, when we are absorbed in objects that are nontemporal or other-temporal, such as structures of thought or fictions of the imagination—moments enough when we go out of time, to allow us to think of ourselves as in some dimension atemporal. "Temporality" is the philosophers' word both for the quality of our time and for its ground. That human beings live in part temporally is beyond question, but whether we have a grounding Temporality seems to me most questionable, since time is not entity enough to have an intelligible essence, and to lever the fluidity and unreality of time into an ontological Temporality seems to me to fabricate philosophical depth by redescription.

Instead we have a temporal form, whose two elements seem to be first, an originary pulsing of the soul together with an ability to count its pulses, and second, a capacity for the orderly storing—and often treasuring—of the unreal residue of perception. These two aspects of our temporality can be represented as orthogonal to each other: horizontal succession-time, and astride of it, vertical phase-time—at once its source and its rider. Where they intersect the ever-advancing pulse of consciousness and the presence of the world before it coincide. The figure is a description, not an explanation. No one knows how human temporality comes to be dual or how well it really accommodates the world's variations.

Above all, no one really knows how the two axes come to collapse into each other in our ordinary, unreflective life-time, so that past and future seem to stretch out behind and before—and with them the world. I have isolated some impossibilities and put forth some conjectures, but when all is said and done, time remains what may well be called a mystery, lucid in parts and obscure as a whole.

(9) Let me end as I began with a glance at the *question itself*. Is the question "What is time?" really much stranger than any other perplexity that might prompt philosophical inquiry?

As poetry takes what is familiar and turns it into something strange and wonderful, so philosophy, which is said to begin in estranged wonder—and is therefore in its first moments quite like poetry—usually ends with familiarizing explanations. For those first wondering questions soon elicit strange and abstruse terms, and these are the ones philosophers go on to explicate, over and over. "Being" and "Nonbeing" are the prime examples that come from the very beginning of philosophy and whose power reaches well into this book.

It really is otherwise with time. By a wild guess, for every time the word "Being" is uttered in daily speech, the word "time" is mentioned a million times. Time is spoken of incessantly and is unutterably familiar. The natural and social world abounds in occasions for the measurement and management of time, but even without much reflection, people do distinguish from these a personal, human time that comes from within. It is not from those cool remote regions of the mind where thought officiates that time seems to come, but from the center of the living soul, our intimate interior. But then, when we are asked to speak about this intimate mode of ours, we are the more stunned by the estrangement, the more familiarly we have lived with time. And even though protracted inquiry may have eventually yielded for us the passing satisfactions of some apt descriptions and the temporary perch of better-framed perplexities, still it has given us no brisk answer to the straight question: What, then, *is* time?

If you have ever descended into the fantastic underworld of a great cavern, you may have seen the reflection of a throng of stalactites hanging over a preternaturally still pool. It appears as an inverse stalagmitic city of steeples and spires rising deep under the deceptively translucent surface of the water. So it is with the observations and the theories that have been the study of this book. Each, upon reflection, reveals its sunken realm of problems and perplexities, deep as the theories were high and as elusive to the grasp. Thus inquiry into time begins in the startled wonder of an answer owed and ends in the settled wonder of a mystery delineated.

Reference Bibliography

Arendt, Hannah. 1961. *Between Past and Future*. New York: Viking Press.

Aveni, Anthony F. 1989. *Empires of Time: Calendars, Clocks, and Cultures*. New York: Basic Books.

Barrow, Isaac. 1670. *Geometrical Lectures*, Lecture I. In *The Concepts of Space and Time*, ed. Milič Čapek. Dortrecht: Reidel, 1976.

Beauregard, Olivier Costa de. 1981. "Time in Relativity Theory: Arguments for a Philosophy of Being." In *The Voices of Time*, ed. J. T. Frazer. Amherst: University of Massachusetts Press.

Bergson, Henri. 1889. *Time and Free Will: An Essay on the Immediate Data of Consciousness*, trans. F. L. Pogson. New York: Harper and Brothers, 1960. Original title: *Essai sur les donnés immédiates de la conscience*.

———. 1896. *Matter and Memory*, trans. Nancy M. Paul and W. Scott Palmer. London: Allen and Unwin, 1911.

Bernet, Rudolf. 1983. "Is the Present Ever Present?: Phenomenology and the Metaphysics of Presence." In *Husserl and Contemporary Thought*, ed. John Sallis. Atlantic Highlands, N.J.: Humanities Press.

Bloch, Maurice. 1996. "Internal and External Memory: Different Ways of Being in History." In *Tense Past: Cultural Essays in Trauma and Memory*, ed. Paul Antze and Michael Lambek. New York: Routledge.

Block, Richard A. 1996. "Memory Systems of the Brain." In *Dimensions of Time and Life: The Study of Time VIII*, ed. J. T. Frazer and M. P. Soulsby. Madison, Conn.: International Universities Press.

Boorstin, Daniel J. 1985. *The Discoverers*. New York: Vintage Books.

Borges, Jorge Luis. 1962. "The Garden of the Forking Paths." In *Ficciones*, ed. Anthony Kerrigan. New York: Grove Press.

———. 1969. *Labyrinths: Selected Stories and Other Writings*, ed. Donald A. Yates and James E. Irby. New York: New Directions.

Boulding, Elise, and Kenneth E. Boulding. 1995. *The Future: Images and Processes*. Thousand Oaks, Calif.: Sage Publications.

Brandt, Anthony. "A Short Natural History of Nostalgia." *Atlantic Monthly* (December 1978):53–63.

Brann, Eva. 1991. *The World of the Imagination: Sum and Substance*. Lanham, Md.: Rowman and Littlefield.

———. 1994. "Plato." In *Encyclopedia of Time*, ed. Samuel L. Macey. New York: Garland.

Brenneis, C. Brooks. 1997. *Recovered Memories of Trauma: Transferring the Present to the Past*. Madison, Conn.: International Universities Press.

Brough, John B. "Husserl on Memory." *The Monist* 59 (January 1975).

———. 1977. "The Emergence of an Absolute Consciousness." In *Husserl: Expositions and Appraisals*, ed. Frederick Elliston and Peter McCormick. Notre Dame: University of Notre Dame Press.

———. 1987. "Temporality and the Presence of Language: Reflections on Husserl's Phenomenology of Time-Consciousness." In *Phenomenology of Temporality: Time and Language*. Simon Silverman Phenomenology Center. Pittsburgh: Duquesne University.

———. 1989. "Husserl's Phenomenology of Time Consciousness." In *Husserl's Phenomenology*, ed. J. N. Mohanty and William McKenna. Lanham, Md.: University Press of America.

Brown, Peter. 1967. *Augustine of Hippo*. Berkeley: University of California Press.

Buckley, James H. 1966. *The Triumph of Time: A Study of Victorian Concepts of Time, History, Progress, and Decadence*. Cambridge: Belknap Press.

Burke, Edmund. 1790. *Reflections on the Revolution in France*. Baltimore: Penguin Books, 1969.

Campbell, Jeremy. 1986. *Winston Churchill's Afternoon Nap: A Wide-Awake Inquiry into the Human Nature of Time*. New York: Simon and Schuster.

Čapek, Milič. 1981. "Time in Relativity Theory: Arguments for a Philosophy of Becoming." In *The Voices of Time*. See Beauregard 1981.

———. 1986. "The Unreality and Indeterminacy of the Future in the Light of Contemporary Physics." In *Physics and the Ultimate Significance of Time*. See Griffin 1986.

Casey, Edward S. 1987. *Remembering: A Phenomenological Study*. Bloomington: Indiana University Press.

Coleman, Janet. 1992. *Ancient and Medieval Memories: Studies in the Reconstruction of the Past*. Cambridge: Cambridge University Press.

Curtius, Ernst. 1948. *Europäische Literatur und Lateinisches Mittelalter*. Bern: Francke Verlag.

Deacon, Terrence W. 1997. *The Symbolic Species: The Co-evolution of Language and the Brain*. New York: W. W. Norton.

Derrida, Jacques. 1939. *Edmund Husserl's Origin of Geometry: An Introduction*, trans. John Leavey. Stony Brook, N.Y.: Hays, 1978.

———. 1968. "*Ousia* and *Gramme*, Note on a Note from *Being and Time*." In *Margins of Philosophy*, trans. Alan Bass. Chicago: University of Chicago Press, 1982.

Dohrn-van Rossum, Gerhard. 1996. *History of the Hour: Clocks and Modern Temporal Orders*, trans. Thomas Dunlap. Chicago: University of Chicago Press.

Dummett, Michael. "Bringing About the Past." In *The Philosophy of Time*. See Le Poidevin and MacBeath 1993.

Durrell, Lawrence. 1960. *Clea*: Book IV of the *Alexandria Quartet*. New York: Pocket Books.

Eliade, Mircea. 1954. *The Myth of Eternal Return, or Cosmos and History*, trans. Willard Trask. Princeton: Princeton University Press.

Fagg, Lawrence W. 1995. *The Becoming of Time: Integrating Physical and Religious Time*. Atlanta: Scholars Press.

Feather, Norman. 1961. *Mass, Length and Time*. Baltimore: Penguin Books.

Feynman, Richard. 1963. *Lectures on Physics*. Reading, Mass.: Addison-Wesley.

Findlay, John. "Husserl's Analysis of the Inner Time-Consciousness." *The Monist* 59 (January 1975):3–20.

Fink, Eugen. 1930. "Vergegenwärtigung und Bild: Beiträge zur Phänomenologie der Unwirklichkeit" ("Presentification and Picture: Contributions to the Phenomenology of Inactuality"). In *Studien zur Phänomenologie*: *1930–1939*. The Hague: Nijhoff, 1966.

Foerster, Eckart. 1989. *Kant's Transcendental Deductions: The Three Critiques and the Opus Posthumum*. Stanford: Stanford University Press.

Frank, P. 1976. "Is the Future Already Here?" In *The Concepts of Space and Time*. See Barrow 1670.

Frazer, J. T. 1987. *Time: The Familiar Stranger.* Amherst: University of Massachusetts Press.

———, ed. 1989. *Time and Mind: Interdisciplinary Issues: The Study of Time VI.* Madison, Conn.: International Universities Press.

Frazer, J. T., N. Lawrence, and F. C. Haber, eds. 1986. *Time, Science, and Society in China and the West. The Study of Time V.* Amherst: University of Massachusetts Press.

Friedman, Michael. 1992. *Kant and the Exact Sciences.* Cambridge: Harvard University Press.

Friedman, William. 1990. *About Time: Inventing the Fourth Dimension*. Cambridge: MIT Press.

Gale, Richard M. 1971. "'Here' and 'Now.'" In *Basic Issues in the Philosophy of Time*, ed. Eugene Freeman and Wilfrid Sellars. La Salle, Ill.: Open Court.

Galgan, Gerald J. 1982. *The Logic of Modernity*. New York: New York University Press.

Galileo. 1638. *Two New Sciences*, trans. Stillman Drake. Madison: University of Wisconsin Press, 1974.

Gallagher, Shaun. 1998. *The Inordinance of Time*. Evanston: Northwestern University Press.

Gelven, Michael. 1970. *A Commentary on Heidegger's "Being and Time."* New York: Harper and Row.

Gilborn, Craig. "Looking at the Coke Bottle." *Museum News* (December 1968):13–18.

Gilson, Étienne. 1961. *The Christian Philosophy of Saint Augustine*. London: Victor Gollancz.

Gould, Stephen J. 1987. *Time's Arrow, Time's Cycle: Myth and Metaphor in the Discovery of Geological Time*. Cambridge: Harvard University Press.

Grant, George.1995. *Time as History*. Toronto: University of Toronto Press.

Griffin, David R., ed. 1986. *Physics and the Ultimate Significance of Time*. Albany: State University of New York Press.

Grudin, Robert. 1982. *Time and the Art of Living*. New York: Ticknor and Fields.

Habermas, Jürgen. 1985. *The Philosophical Discourse of Modernity: Twelve Lectures* (1985), trans. Frederick Lawrence. Cambridge: MIT Press.

Halbwachs, Maurice. 1992. *On Collective Memory*, trans. Lewis A. Coser. Chicago: University of Chicago Press.

Hall, Edward T. 1959. *The Silent Language*. Greenwich, Conn.: Fawcett.

Harris, Errol E. 1988. *The Reality of Time*. Albany, New York: State University of New York Press.

Hawking, Stephen W. 1988. *A Brief History of Time*. New York: Bantam Books.

Heidegger, Martin. 1924. *The Concept of Time*, trans. William McNeil. Oxford: Blackwell, 1992.

———. 1926. *Logik: Die Frage nach der Wahrheit.* Frankfurt: Vittorio Klostermann, 1976.

———. 1979. *History of the Concept of Time*, trans. Theodore Kisiel. Bloomington: Indiana University Press, 1984.

Helm, Bertrand P. 1985. *Time and Reality in American Philosophy*. Amherst: University of Massachusetts Press.

Herz, Heinrich. 1894. *Principles of Mechanics*. New York: Dover, 1956.

Heschel, Abraham J. 1951. *The Sabbath: Its Meaning for Modern Man*. New York: Farrar, Strauss, and Giroux, 1996.

Hintikka, Jaako. 1973. *Time and Necessity: Studies in Aristotle's Theory of Modality*. Oxford: Clarendon Press.

Hoffman, Piotr. 1986. *Doubt, Time, Violence*. Chicago: University of Chicago Press.

Husserl, Edmund. 1966. *Husserliana*: *Edmund Husserl, Gesammelte Werke X: Zur Phänomenologie des inneren Zeitbewusstseins, 1893–1917*, ed. Rudolf Boehm. The Hague: Nijhoff.

———. 1991. *Edmund Husserl: On the Phenomenology of the Consciousness of Internal Time (1893–1917)*, trans. John B. Brough. In *Edmund Husserl: Collected Works* IV, ed. Rudolf Bernet. Dortrecht: Kluwer.

James, William. 1892. *Psychology*. Cleveland: World, 1948.

Jesperson, Otto. 1924. *The Philosophy of Grammar*. London: Allen and Unwin.

Johnson, Mark. 1987. *The Body in the Mind: The Bodily Basis of Meaning, Imagination, and Reason*. Chicago: University of Chicago Press.

Jouvenel, Bertrand de. 1967. *The Art of Conjecture*, trans. Nikita Lary. New York: Basic Books.

Kahn, Charles H. "The Greek Verb 'To Be' and the Concept of Being." *Foundations of Language*, 2 (August 1966):245–65.

Kahn, Herman, and Anthony J. Wiener. 1967. *The Year 2000: A Framework for Speculation on the Next Thirty-three Years*. New York: Macmillan.

Kantorowicz, Ernst H. 1957. *The King's Two Bodies: A Study in Medieval Political Theology*. Princeton: Princeton University Press.

Kellert, Stephen H. 1993. *In the Wake of Chaos: Unpredictable Order in Dynamical Systems*. Chicago: University of Chicago Press.

Kepler, Johannes. 1609. *New Astronomy*, trans. William H. Donahue. Cambridge: Cambridge University Press, 1992.

Kermode, Frank. 1966. *The Sense of an Ending: Studies in the Theory of Fiction*. Oxford: Oxford University Press.

Kilwardby, Robert. Late 1250s. *De Tempore, de Spiritu Fantastico (On Time and Imagination)*, ed. P. Osmund Lewry, O.P. Oxford: Oxford University Press, 1987.

Klatzky, Roberta L. 1980. *Human Memory: Structures and Processes*. San Francisco: Freeman.

Klein, Jacob. 1934. *Greek Mathematical Thought and the Origin of Algebra*, trans. Eva Brann. N.Y.: Dover, 1968.

Kockelmans, Joseph. 1967. "What Is Phenomenology? The Philosophy of Edmund Husserl and Its Interpretation." In *Phenomenology*, ed. Joseph Kockelmans. Garden City, N.Y.: Doubleday.

Kohák, Erazim. 1978. *Idea and Experience: Edmund Husserl's Project of Phenomenology in Ideas I*. Chicago: University of Chicago Press.

Kosselleck, Reinhart. 1985. *Futures Past: On the Semantics of Historical Time*, trans. Keith Tribe. Cambridge: MIT Press.

Kubler, George. 1962. *The Shape of Time: Remarks on the History of Things*. New Haven: Yale University Press.

Le Poidevin, Robin. 1993. "Relationism and Temporal Topology: Physics or Metaphysics?" In *The Philosophy of Time*. See Le Poidevin and MacBeath 1993.

Le Poidevin, Robin, and Murray MacBeath, eds. 1993. *The Philosophy of Time*, Part I. Oxford: Oxford University Press.

Lears, Jackson. "Looking Backward: In Defense of Nostalgia." *Lingua Franca* (January 1998):59–66.

Leibniz, Gottfried Wilhelm. c. 1714. "Metaphysical Foundations of Mathematics." In *Philosophical Papers and Letters II*, ed. Leroy E. Loemker. Chicago: University of Chicago Press, 1956.

Léon-Portilla, Miguel. 1988. *Time and Reality in the Thought of the Maya*. 2d ed. Norman: University of Oklahoma Press.

Lieb, Irwin C. 1991. *Past, Present, and Future: A Philosophical Essay about Time*. Urbana: University of Illinois Press.

Lowenthal, David. 1985. *The Past Is a Foreign Country*. Cambridge: Cambridge University Press.

Lyons, John. 1977. *Semantics 2*. Cambridge: Cambridge University Press.

Mach, Ernst. 1883. *Science of Mechanics: A Critical and Historical Account of Its Development*. La Salle, Ill.: Open Court, 1960.

MacRae, Cathi Dunn. 1998. "Time Fantasy: Then and Now." Chap. 7 in *Presenting Young Adult Fantasy Fiction*. New York: Twayne.

Makreel, Rudolf. 1990. *Imagination and Interpretation in Kant: The Hermeneutical Import of the Critique of Judgment*. Chicago: University of Chicago Press.

Mandelstam, Nadezhda. 1972. *Hope Abandoned*, trans. Max Hayward. New York: Atheneum, 1981.

Marcey, Samuel L., ed. 1991. *Time: A Bibliographic Guide*. New York: Garland.

Marx, Karl. 1867. *Capital: A Critique of Political Economy*. New York: The Modern Library.

———. 1964. *The Economic and Philosophical Manuscripts of 1844*, ed. Dirk J. Struik. New York: International Publishers.

McInerney, Peter. 1991. *Time and Experience*. Philadelphia: Temple University Press.

McTaggart, J. M. E. 1927. *The Nature of Existence II*. Cambridge: Cambridge University Press.

Mead, George Herbert. 1930. *The Philosophy of the Present*. La Salle, Ill.: Open Court, 1959.

Merleau-Ponty, M. 1962. *Phenomenology of Perception*, trans. Colin Smith. London: Routledge and Kegan Paul.

Michon, John A. 1990. "Implicit and Explicit Representations of Time." In *Cognitive Models of Psychological Time*, ed. Richard A. Block. Hillsdale, N.J.: Erlbaum.

Miller, G. A. "The Magical Number Seven, Plus or Minus Two: Some Limits on Our Capacity for Processing Information." *Psychological Review* 63 (1956):81–97.

Mo, Suchoon S. 1990. "Time Reversal and Lateral Reversal." In *Cognitive Models of Psychological Time*. See Michon 1990.

Morris, Richard. 1985. *Time's Arrows: Scientific Attitudes toward Time*. New York: Simon and Schuster.

Newton-Smith, W. H. 1980. *The Structure of Time*. London: Routledge and Kegan Paul.

Nowotny, Helga. 1989. "Mind, Technologies, and Collective Time Consciousness." In *Time and Mind*. See Frazer 1989.

Oaklander, L. Nathan, and Quentin Smith, eds. 1994. *The New Theory of Time*. Part I. New Haven: Yale University Press.

O'Rahilly, Alfred. 1938. *Electromagnetic Theory: A Critical Examination of Fundamentals*. Vol. 2. New York: Dover, 1965.

Panofsky, Erwin. 1962. *Studies in Iconology: Humanistic Themes in the Art of the Renaissance*. New York: Harper and Row.

Patton, H. J. 1936. *Kant's Metaphysic of Experience: A Commentary on the First Half of the Kritik der reinen Vernunft.* New York: Macmillan.

Poeppel, Ernst. 1988. *Mindworks: Time and Conscious Experience*, trans. Tom Artin. San Diego: Harcourt Brace Jovanovich.

Resnick, Robert, and David Halliday. 1966. *Physics I.* New York: Wiley.

Ricoeur, Paul. 1967. *Husserl: An Analysis of His Phenomenology*, trans. Edward G. Ballard and Lester E. Embree. Evanston: Northwestern University Press.

———. 1984. *Time and Narrative*. Vol. 1, trans. Kathleen McLaughlin and David Pellauer. Chicago: University of Chicago Press.

———. 1988. *Time and Narrative*. Vol. 3, trans. Kathleen Blamey and David Pellauer. Chicago: University of Chicago Press.

Rifkin, Jeremy. 1987. *Time Wars: The Primary Conflict in Human History*. New York: Simon and Schuster.

Rosen, Stanley. 1969. *Nihilism: A Philosophical Essay.* New Haven: Yale University Press.

Sacks, Oliver. 1995. "The Landscape of His Dreams." In *An Anthropologist on Mars: Seven Paradoxical Tales*. New York: Knopf.

Sambursky, S., and S. Pines. 1971. *The Concept of Time in Late Neoplatonism*. Jerusalem: Israel Academy of Sciences and Humanities.

Sartre, Jean-Paul. 1943. *Being and Nothingness: A Phenomenological Essay on Ontology*. New York: Pocket Books, 1956.

Savitt, Steven F. 1995. "Introduction: Becoming." In *Time's Arrows Today: Recent Physical and Philosophical Work on the Direction of Time*. Cambridge: Cambridge University Press.

Schacter, Daniel L. 1996. *Searching for Memory, the Brain, the Mind, and the Past*. New York: Basic Books.

Schama, Simon. 1995. *Landscape and Memory*. New York: Knopf.

Schilpp, Paul A., ed. 1949. *Albert Einstein: Philosopher-Scientist*. Vol. I. La Salle, Ill.: Open Court. Publishers.

Schnapp, Alain. 1997. *The Discovery of the Past*. New York: Abrams.

Schneider, Ulrich. 1990. *Die Vergangenheit des Geistes: Eine Archäologie der Philosophiegeschichte*. Frankfurt: Suhrkamp.

Sherover, Charles M. 1971. *Heidegger, Kant and Time*. Bloomington: University of Indiana Press.

Shoemaker, Sidney. 1993. "Time Without Change." In *The Philosophy of Time*. See Le Poidevin and MacBeath 1993.

Smith, Brian. 1966. *Memory*. London: Allen and Unwin.

Smith, Quentin. 1994. "Introduction: The Old and the New Theory of Time." In *The New Theory of Time*. See Oaklander and Smith 1994.

Sobel, Dava. 1995. *Longitude*. New York: Penguin Books.

Sorabji, Richard. 1983. *Time, Creation and the Continuum: Theories in Antiquity and the Early Middle Ages*. Ithaca: Cornell University Press.

Taylor, A. E. 1928. *Commentary on Plato's Timaeus*. Oxford: Clarendon Press.

Taylor, E., and J. Wheeler. 1963. *Spacetime Physics*. San Francisco: Freeman.

Thompson, E. P. 1991. *Customs in Common*. New York: New Press.

Thomsen, Christian W., and Hans Holländer, eds. 1984. *Augenblick und Zeitpunkt: Studien zur Zeitstruktur und Zeitmetaphorik in Kunst und Wissenschaften*. Darmstadt: Wissenschaftliche Buchgesellschaft.

Tooley, Michael. 1997. *Time, Tense, and Causation*. Oxford: Clarendon Press.

Watanabi, Satosi. 1981. "Time and the Probabilistic View of the World." In *The Voices of Time*. See Beauregard 1981.

Whitehead, Alfred North. 1927. *Symbolism: Its Meaning and Effect*. New York: Fordham University Press, 1985.

Whitrow, G. J. 1961. *The Natural Philosophy of Time*. New York: Harper and Row.

———. 1988. *Time in History: The Evolution of Our General Awareness of Time and Temporal Perspective*. Oxford: Oxford University Press.

Wilcox, Donald J. 1987. *The Measure of Times Past: Pre-Newtonian Chronologies and the Rhetoric of Relative Time*. Chicago: University of Chicago Press.

Yerushalmi, Yosef H. 1989. *Zakhor: Jewish History and Jewish Memory*. New York: Schocken Books.

Yourgrau, Palle. 1991. *The Disappearance of Time: Kurt Goedel and the Idealistic Tradition in Philosophy*. Cambridge: Cambridge University Press.

Zerubavel, Eviatar. 1985. *The Seven Day Circle: The History and Meaning of the Week*. Chicago: University of Chicago Press.

Index

C

D

E

H

I

N

R

U

V

W

Y

About the Author

Eva Brann began her working life at the American excavations of the Athenian marketplace, studying a particularly exuberant period in early Attic pottery. In 1957 she left archaeology to become a tutor (the only title bestowed) at St. John's College in Annapolis, Maryland, where she has taught ever since, except for seven years' service as dean (1990–97). St. John's has an all-required liberal arts curriculum, including language, mathematics, science, and discussion seminars in great books of literature and philosophy. Since every tutor is expected to teach what all the students are asked to learn, teaching at the college means strenuous life-long learning, the kind that stimulates inquiry into central human issues such as the imagination and time. These two elusive and enchanting preoccupations resulted in *The World of the Imagination* and *What, Then, Is Time*—both published by Rowman & Littlefield. A future book, in what has turned out to be something like a trilogy, will pursue the human capacity haunting both previous books, that of living vividly with absent presences, lapsing presents, and all sorts of other Nonbeing.